MEMOIRS
of the
American Mathematical Society

Number 1376

Intrinsic Approach to Galois Theory of q-Difference Equations

Lucia Di Vizio
Charlotte Hardouin

With a Preface to Part 4 by Anne Granier

Library of Congress Cataloging-in-Publication Data

Cataloging-in-Publication Data has been applied for by the AMS.
See http://www.loc.gov/publish/cip/.
DOI: https://doi.org/10.1090/memo/1376

Memoirs of the American Mathematical Society

This journal is devoted entirely to research in pure and applied mathematics.

Subscription information. Beginning with the January 2010 issue, *Memoirs* is accessible from www.ams.org/journals. The 2022 subscription begins with volume 275 and consists of six mailings, each containing one or more numbers. Subscription prices for 2022 are as follows: for paper delivery, US$1085 list, US$868 institutional member; for electronic delivery, US$955 list, US$764 institutional member. Upon request, subscribers to paper delivery of this journal are also entitled to receive electronic delivery. If ordering the paper version, add US$22 for delivery within the United States; US$85 for outside the United States. Subscription renewals are subject to late fees. See www.ams.org/help-faq for more journal subscription information. Each number may be ordered separately; *please specify number* when ordering an individual number.

Back number information. For back issues see www.ams.org/backvols.

Subscriptions and orders should be addressed to the American Mathematical Society, P. O. Box 845904, Boston, MA 02284-5904 USA. *All orders must be accompanied by payment.* Other correspondence should be addressed to 201 Charles Street, Providence, RI 02904-2213 USA.

Copying and reprinting. Individual readers of this publication, and nonprofit libraries acting for them, are permitted to make fair use of the material, such as to copy select pages for use in teaching or research. Permission is granted to quote brief passages from this publication in reviews, provided the customary acknowledgment of the source is given.

Republication, systematic copying, or multiple reproduction of any material in this publication is permitted only under license from the American Mathematical Society. Requests for permission to reuse portions of AMS publication content are handled by the Copyright Clearance Center. For more information, please visit www.ams.org/publications/pubpermissions.

Send requests for translation rights and licensed reprints to reprint-permission@ams.org.

Excluded from these provisions is material for which the author holds copyright. In such cases, requests for permission to reuse or reprint material should be addressed directly to the author(s). Copyright ownership is indicated on the copyright page, or on the lower right-hand corner of the first page of each article within proceedings volumes.

Memoirs of the American Mathematical Society (ISSN 0065-9266 (print); 1947-6221 (online)) is published bimonthly (each volume consisting usually of more than one number) by the American Mathematical Society at 201 Charles Street, Providence, RI 02904-2213 USA. Periodicals postage paid at Providence, RI. Postmaster: Send address changes to Memoirs, American Mathematical Society, 201 Charles Street, Providence, RI 02904-2213 USA.

©2022 Lucia Di Vizio, Charlotte Hardouin, and Anne Granier. All rights reserved.
This publication is indexed in *Mathematical Reviews*®, *Zentralblatt MATH*, *Science Citation Index*®, *Science Citation Index*^{TM}*-Expanded*, *ISI Alerting Services*^{SM}, *SciSearch*®, *Research Alert*®, *CompuMath Citation Index*®, *Current Contents*®/*Physical, Chemical & Earth Sciences*.
This publication is archived in *Portico* and *CLOCKSS*.
Printed in the United States of America.

∞ The paper used in this book is acid-free and falls within the guidelines established to ensure permanence and durability.
Visit the AMS home page at https://www.ams.org/

10 9 8 7 6 5 4 3 2 1   27 26 25 24 23 22

Contents

Introduction	vii
Grothendieck conjecture for q-difference equations	ix
Intrinsic Galois groups	x
Comparison with Malgrange-Granier Galois theory for non-linear differential equations	xi
Acknowledgments	xii
Part 1. Introduction to q-difference equations	1
Chapter 1. Generalities on q-difference modules	3
1.1. Basic definitions	3
1.2. q-difference modules, systems and equations	6
1.3. Some remarks on solutions	7
1.4. Trivial q-difference modules	8
Chapter 2. Formal classification of singularities	11
2.1. Regularity	11
2.2. Irregularity	12
Part 2. Triviality of q-difference equations with rational coefficients	13
Chapter 3. Rationality of solutions, when q is an algebraic number	15
3.1. The case of q algebraic, not a root of unity	15
3.2. Global nilpotence.	18
3.3. Proof of Theorem 3.8 (and of Theorem 3.6)	19
Chapter 4. Rationality of solutions when q is transcendental	23
4.1. Statement of the main result	23
4.2. Regularity and triviality of the exponents	24
4.3. Proof of Theorem 4.2	28
4.4. Link with iterative q-difference equations	29
Chapter 5. A unified statement	31
Part 3. Intrinsic Galois groups	33
Chapter 6. The intrinsic Galois group	35
6.1. Definition and first properties	35
6.2. Arithmetic characterization of the intrinsic Galois group	36
6.3. Finite intrinsic Galois groups	37
6.4. Intrinsic Galois group of a q-difference module over $\mathbb{C}(x)$, for $q \neq 0, 1$	38

Chapter 7. The parametrized intrinsic Galois group — 41
7.1. Differential and difference algebra — 41
7.2. Parametrized intrinsic Galois groups — 42
7.3. Characterization of the parametrized intrinsic Galois group by curvatures — 45
7.4. Parametrized intrinsic Galois group of a q-difference module over $\mathbb{C}(x)$, for $q \neq 0, 1$ — 47
7.5. The example of the Jacobi Theta function — 47

Part 4. Comparison with the non-linear theory — 49

Chapter 8. Preface to Part 4. The Galois D-groupoid of a q-difference system, by Anne Granier — 51
8.1. Definitions — 51
8.2. A bound for the Galois D-groupoid of a linear q-difference system — 52
8.3. Groups from the Galois D-groupoid of a linear q-difference system — 54

Chapter 9. Comparison of the parametrized intrinsic Galois group with the Galois D-groupoid — 57
9.1. The Kolchin closure of the Dynamics and the Malgrange-Granier groupoid — 58
9.2. The groupoid $\mathcal{G}al^{alg}(A(x))$ — 58
9.3. The Galois D-groupoid $\mathcal{G}al(A(x))$ vs the intrinsic parametrized Galois group — 62
9.4. Comparison with known results — 64

Bibliography — 67

Abstract

The Galois theory of difference equations has witnessed a major evolution in the last two decades. In the particular case of q-difference equations, authors have introduced several different Galois theories. In this memoir we consider an arithmetic approach to the Galois theory of q-difference equations and we use it to establish an arithmetical description of some of the Galois groups attached to q-difference systems.

Received by the editor May 6, 2014, and, in revised form, October 5, 2017, and November 12, 2018.

Article electronically published on August 3, 2022.

DOI: `https://doi.org/10.1090/memo/1376`

2020 *Mathematics Subject Classification.* Primary: 39A13, 12H10.

Key words and phrases. Generic Galois group; intrinsic Galois group; q-difference equations; differential Tannakian categories; Kolchin differential groups; Grothendieck conjecture on p-curvatures; D-groupoid.

The first author is affiliated with the Laboratoire de Mathématiques UMR 8100, CNRS, Université de Versailles-St Quentin, 45 avenue des États-Unis 78035 Versailles cedex, France. Email: divizio@math.cnrs.fr.

The second author is affiliated with the Institut de Mathématiques de Toulouse, 118 route de Narbonne, 31062 Toulouse Cedex 9, France. Email: hardouin@math.univ-toulouse.fr.

The aurhor of the preface to Part 4 is affiliated with the Institut de Mathématiques de Toulouse, 118 route de Narbonne, 31062 Toulouse Cedex 9, France.

Work partially supported by ANR-06-JCJC-0028 and ECOS-Nord C12M01.

©2022 Lucia Di Vizio, Charlotte Hardouin, and Anne Granier

Introduction

The Galois theory of difference equations has witnessed a major evolution in the last two decades. In the particular case of q-difference equations, authors have introduced several different Galois theories. In this memoir we consider an arithmetic approach to the Galois theory of q-difference equations and we use it to establish an arithmetical description of some of the Galois groups attached to q-difference systems.

Let q be a non-zero element of the field $\mathbb{C}$ of complex numbers. A (linear) q-difference system is a functional equation of the form

$$(0.1) \qquad Y(qx) = A(x)Y(x), \text{ with } A(x) \in \mathrm{GL}_\nu(\mathbb{C}(x)).$$

The *leitmotif* of the paper is the Galoisian properties of the so-called dynamics of the system (0.1), namely the set of maps obtained by iteration of the map $(x, X) \longmapsto (qx, A(x)X)$, defined over $U \times \mathbb{C}^\nu$, where U is an open subset of $\mathbb{P}^1_\mathbb{C}$.

Theorem 5.1 below proves that the algebraic nature of the solutions of the q-difference system (0.1) is entirely determined by the specialization of certain subsequences of the dynamics $\bigl(A(q^{n-1}x)\ldots A(x)\bigr)_{n\in\mathbb{N}}$, that are called the curvatures of the q-difference system. This Theorem extends the main result of [**DV02**], in which the assumption that K is a number field, and hence that q is algebraic, is crucial. Here we only assume K to be a finitely generated $\mathbb{Q}$-algebra and q can be any number, algebraic or transcendental. We state here Theorem 5.1 in the particular case $K = \mathbb{Q}(q)$ and under the assumption that q is a transcendental number:

THEOREM 1. *Let $A(x) \in \mathrm{GL}_\nu(\mathbb{Q}(q,x))$. The q-difference system $Y(qx) = A(x)Y(x)$ admits a full set of solutions in $\mathbb{Q}(q,x)$ if and only if for almost all $n \in \mathbb{N}$ there exists a n-th primitive root of unity ζ_n such that $A(q^{n-1}x)\ldots A(x)$ specializes to the identity matrix at $q = \zeta_n$.*

One could ask whether an analogous statement holds for more general difference equations, that is for instance for dynamics induced by the action an algebraic group. It appears that our statement fails to be true already when one replaces the multiplicative group by the additive group which corresponds to the case of finite difference equations, i.e., for equations associated to the operator $x \mapsto x+1$. In [**vdPS97**, page 58, §5.4], the authors provide a counterexample.

Relying on the above rationality criteria, one is able to provide an arithmetic set of generators for some of the Galois groups attached to q-difference systems. More precisely, in [**HS08**], the authors attach to a q-difference system a linear differential algebraic group *à la Kolchin*, that is a group of matrices defined as the set of zeros of a finite number of algebraic differential equations. The defining equations of this parametrized Galois group encode the differential algebraic relations satisfied by the solutions of the q-difference system. The work of Hardouin and Singer generalizes the *classical* Picard-Vessiot theory of linear difference systems as developed in [**vdPS97**] but requires that the field of σ_q-constants is differentially closed (see 7.1). We attach to a q-difference system $Y(qx) = A(x)Y(x)$, with $A(x) \in \mathrm{GL}_n(K(x))$, a differential algebraic group scheme, that we call parametrized intrinsic Galois group. Roughly, this differential algebraic group scheme is linked to the differential algebraic relations satisfied by the entries of $A(x)$, in the sense that it only relies on constructions of differential algebra of the associated q-difference module, and therefore on the associated matrix constructions of $A(x)$ and its dynamics. The advantages of considering this group are its intrinsic nature and its arithmetic description (see Chapter 7), which is an analogue of the conjectural description obtained by Katz in [**Kat82**] for the Lie algebra of the intrinsic Galois group of a linear differential system. One can show that above a suitable differential field extension of $\mathbb{C}(x)$ the parametrized Galois group and its intrinsic version become isomorphic. This allows to give an arithmetical description of the parametrized Galois group, that might be suitable to build computation algorithms. Indeed, unlike the case of linear differential systems, the computation of the curvatures of a q-difference system relies only on matrix multiplication. Thus, one may hope to develop fast algorithms to compute the curvatures and perhaps also the parametrized intrinsic Galois group in terms of differential polynomial equations annihilated by the curvatures. See [**BS09**] in the differential case.[1] Notice that the arithmetic description of the parametrized intrinsic Galois group provides an arithmetic answer to the problem of the rationality of solutions of q-difference systems as well as the control of their differential dependencies with respect to parameters (see for instance [**AR13**] for some algorithms that tackle these questions).

Finally, the description of the parametrized intrinsic Galois group in terms of curvatures allows us to understand the link between the linear and non-linear Galois theory of q-difference systems. In [**Gra12**], A. Granier introduces a Galois D-groupoid for non-linear q-difference equations, in the spirit of Malgrange's work. In Corollary 9.12, we show, using once more the curvature characterization of the parametrized intrinsic Galois group, that the Malgrange-Granier D-groupoid generalizes the parametrized intrinsic Galois group to the non-linear case. Thanks to our comparison results, we are able to compare the Malgrange-Granier D-groupoid to the parametrized Galois group of Hardouin-Singer. This answers a question of Malgrange ([**Mal09**], page 2]) on the relation among D-groupoids and Kolchin's differential algebraic groups.

Description of the main results

The paper being relatively long, we give here a quite detailed description of the content. Part 1 is an introduction to q-difference equations and explains some preliminary results.

---

[1]One should also cite [**BCS14**,**BCS15**,**BCS16**], which have appeared since the submission of this memoir. We should also point out that an algorithm to calculate the Lie algebra of the Galois group of a linear differential equation has been published in [**BCDVW16**]. It can be accelerated using the easy known implication of the Grothendieck-Katz conjecture on p-curvatures. In contrast with previous algorithms, for whom there was no hope of actual implementation, this one is implemented in MAPLE. An analogous algorithm for q-difference equations is a work in progress: T. Dreyfus and M. Poulet have obtained results in this direction.

INTRODUCTION

Grothendieck conjecture for q-difference equations

In [**DV02**], the first author proved a q-difference analogue of the Grothendieck conjecture on p-curvatures, under the assumption that q is an algebraic number and that the field of constants is a number field. In this paper, we generalize this result in two different directions.

Consider a field of rational functions $K(x)$, a transcendental element $q \in K$, such that K is itself a field of rational functions in q of the form $k(q)$, and a q-difference system $Y(qx) = A(x)Y(x)$, with $A(x) \in \mathrm{GL}(K(x))$. We prove the following result (see Theorem 4.2 for a more general and intrinsic result):

THEOREM 2. *A q-difference system $Y(qx) = A(x)Y(x)$, with $A(x) \in \mathrm{GL}_\nu(K(x))$, has a solution matrix in $\mathrm{GL}_\nu(K(x))$ if and only if for almost all positive integer n there exists a primitive n-th root of unity ζ_n such that*

$$\left[A(q^{n-1}x) \cdots A(qx)A(x)\right]_{q=\zeta_n} = 1,$$

where 1 stands for the identity matrix of size ν.

In the present article we work under more general assumptions. Namely, we assume that k is a perfect field, of any characteristic, and that K is a finite extension of $k(q)$. Replacing k by its perfect closure, the theorem above covers all the possible cases in which q is transcendental over the prime field.

Suppose now that q is algebraic over the prime field, and that the characteristic of K is zero. We consider again the q-difference system $Y(qx) = A(x)Y(x)$, with $A(x) \in \mathrm{GL}(K(x))$. We can always suppose that K is actually finitely generated over $\mathbb{Q}$. For the sake of simplicity, we assume in this introduction that $K = \mathbb{Q}(\alpha)$ is a purely transcendental extension and that $q \in \mathbb{Q}$, $q \neq 0, 1, -1$. For almost all rational primes p the image of q in $\mathbb{F}_p$ is well-defined and non-zero, so that there exists a minimal positive number κ_p such that $q^{\kappa_p} \equiv 1$ modulo p. Let ℓ_p be a positive integer such that $1 - q^{\kappa_p} = p^{\ell_p} \frac{h}{g}$, with $h, g \in \mathbb{Z}$ prime to p. We have (see Theorem 3.6):

THEOREM 3. *A q-difference system $Y(qx) = A(x)Y(x)$, with $A(x) \in \mathrm{GL}_\nu(K(x))$, has a solution matrix in $\mathrm{GL}_\nu(K(x))$ if and only if for almost all prime p we have*

$$A(q^{\kappa_p - 1}x) \cdots A(qx)A(x) \equiv 1 \ \text{modulo} \ p^{\ell_p}.$$

The statement above is a little bit imprecise, since we should have introduced a $\mathbb{Z}$-algebra contained in $K(x)$ that would have given a precise sense to the reduction modulo p^{ℓ_p}, for almost all p. The reader will find a more formal statement in Part 2, where the result above is proved under the assumption that K is any finitely generated extension of $\mathbb{Q}$ and that q is an algebraic number, not a root of unity. As already pointed out, the first author proves in [**DV02**, Thm.7.1.1] the statement above under the assumption that K is a number field. Our proof relies on [**DV02**, Thm.7.1.1], in the sense that we consider a transcendence basis of K over $\mathbb{Q}$ as a set of parameters varying in the algebraic closure of $\mathbb{Q}$ and therefore we make a non-trivial reduction to the situation considered in [**DV02**], for sufficiently many special values of the parameters.

Notice that if we start with a q-difference system over $\mathbb{C}(x)$ and a complex number q, which is not a root of unity, then we can always assume, without loss of generality, that we are in one of the two situations above.

The rest of the paper relies on Theorem 2 and Theorem 3 above, in the sense that we first give a geometric equivalent statement of such theorems and then use it to establish a link with the non-linear theory.

Intrinsic Galois groups

Let K be a field of characteristic zero and q a non-zero element of K, which is not a root of unity. We denote by σ_q the q-difference operator $f(x) \mapsto f(qx)$. A q-difference module $\mathcal{M}_{K(x)} = (M_{K(x)}, \Sigma_q)$ over $K(x)$ is a $K(x)$-vector space of finite dimension ν equipped with a σ_q-semilinear bijective operator Σ_q:

$$\Sigma_q(fm) = \sigma_q(f)\Sigma_q(m), \text{ for any } m \in M_{K(x)} \text{ and } f \in K(x).$$

In a given basis of $M_{K(x)}$, the vector of coordinates of an element fixed by Σ_q is solution column of a linear q-difference system of the form

$$(\mathcal{S}_q) \qquad Y(qx) = A(x)Y(x), \text{ with } A(x) \in \mathrm{GL}_\nu(K(x)).$$

We consider the collection $Constr(\mathcal{M}_{K(x)})$ of constructions of linear algebra of $\mathcal{M}_{K(x)}$ (direct sums, tensor products, symmetric and antisymmetric products, duals). The operator Σ_q induces a q-difference operator on every element of $Constr(\mathcal{M}_{K(x)})$, that we still call Σ_q. Then the intrinsic Galois group of $\mathcal{M}_{K(x)}$ is defined as:

$$Gal(\mathcal{M}_{K(x)}) = \{\varphi \in \mathrm{GL}(M_{K(x)}) : \varphi \text{ stabilizes every } \Sigma_q\text{-stable subset}$$

$$\text{in any construction of linear algebra of } M_{K(x)}\}.$$

This definition has to be understood in a functorial way, which allows to endow the intrinsic Galois group with a structure of group scheme over $K(x)$. Of course, this is linked to a Tannakian definition of $Gal(\mathcal{M}_{K(x)})$. As in [**Kat82**], Theorem 2 and Theorem 3 can be reformulated as an arithmetic description of the intrinsic Galois group:

THEOREM 4. *In the notation of Theorem 2 (resp. Theorem 3), the intrinsic Galois group $Gal(\mathcal{M}_{K(x)})$ is the smallest algebraic subgroup of $\mathrm{GL}(M_{K(x)})$, whose specialization at ζ_n contains the specialization of the operator Σ_q^n at ζ_n, for almost all positive integer n and for a choice of a primitive n-th root of unity ζ_n (resp. whose reduction modulo p^{ℓ_p} contains the reduction of the operator $\Sigma_q^{\kappa_p}$ modulo p^{ℓ_p}, for almost all prime p).*

This statement is a little bit informal. The reader will find a precise statement in Chapter 6.

As the notion of intrinsic Galois group is deeply related to the notion of Tannakian category, the notion of parametrized intrinsic Galois group is related to the notion of differential Tannakian category developed in [**Ovc09a**] and [**Kam10**]. We show in this paper how the category of q-difference modules over $K(x)$, equipped with a derivation ∂ commuting with σ_q, such as $x\frac{d}{dx}$, can be endowed with a prolongation functor F and thus turns out to be a differential Tannakian category. Intuitively, if $\mathcal{M}$ is a q-difference module, associated with a q-difference system $\sigma_q(Y) = AY$, the q-difference module $F(\mathcal{M})$ is attached to the q-difference system

$$\sigma_q(Z) = \begin{pmatrix} A & \partial A \\ 0 & A \end{pmatrix} Z.$$

Notice that if Y verifies $\sigma_q(Y) = AY$, then $Z = \begin{pmatrix} Y & \partial(Y) \\ 0 & Y \end{pmatrix}$ is solution of the system above. We consider the family $Constr^\partial(\mathcal{M}_K(x))$ of constructions of differential algebra of $\mathcal{M}_{K(x)}$, that is the smallest family containing $\mathcal{M}_{K(x)}$ and closed with respect to all constructions of linear algebra (direct sums, tensor products, symmetric and antisymmetric products, duals) plus the prolongation functor F. Then the parametrized intrinsic Galois group of $\mathcal{M}_{K(x)}$ is defined as:

$$Gal^\partial(\mathcal{M}_{K(x)}) = \{\varphi \in \mathrm{GL}(M_{K(x)}) : \varphi \text{ stabilizes every } \Sigma_q\text{-stable subset}$$

in any construction of differential algebra of $M_{K(x)}\}$.

The group $Gal^\partial(\mathcal{M}_{K(x)})$ is endowed with a structure of linear differential algebraic group (*cf.* [**Kol73**]). Theorem 2 and Theorem 3 can be reformulated as an arithmetic description of the parametrized intrinsic Galois group:

THEOREM 5. *In the notation of Theorem 2 (resp. Theorem 3), the parametrized intrinsic Galois group $Gal^\partial(\mathcal{M}_{K(x)})$ is the smallest differential algebraic subgroup of $\mathrm{GL}(M_{K(x)})$, whose specialization at ζ_n contains the specialization of the operator Σ_q^n at ζ_n, for almost all positive integer n and for a choice of a primitive n-th root of unity ζ_n (resp. whose reduction modulo p^{ℓ_p} contains the reduction of the operator $\Sigma_q^{\kappa_p}$ modulo p^{ℓ_p}, for almost all prime p).*

Comparison with Malgrange-Granier Galois theory for non-linear differential equations

A. Granier has defined a Galois D-groupoid for nonlinear q-difference equations, in the wake of Malgrange's and Casale's work. In the particular case of a linear system $Y(qx) = A(x)Y(x)$, with $A(x) \in \mathrm{GL}_\nu(\mathbb{C}(x))$, the Malgrange-Granier D-groupoid is the D-envelop of the dynamics, i.e., it encodes all the partial differential equations over $\mathbb{P}^1_\mathbb{C} \times \mathbb{C}^\nu$ with analytic coefficients, satisfied by local diffeomorphisms of the form $(x, X) \mapsto (q^k x, A_k(x)X)$ for all $k \in \mathbb{Z}$, where $A_k(x) \in \mathrm{GL}_\nu(\mathbb{C}(x))$ is the matrix obtained by iterating the system $Y(qx) = A(x)Y(x)$ so that:

$$Y(q^k x) = A_k(x)Y(x).$$

Notice that:
$A_k(x) := A(q^{k-1}x) \ldots A(qx)A(x)$ for all $k \in \mathbb{Z}$, $k > 0$;
$A_0(x) = Id_\nu$;
$A_k(x) := A(q^k x)^{-1} A(q^{k+1}x)^{-1} \ldots A(q^{-1}x)^{-1}$ for all $k \in \mathbb{Z}$, $k < 0$.

Using Theorem 10, we relate this analytic D-groupoid with the more algebraic notion of parametrized intrinsic Galois group. We prove that the solutions in a neighborhood of $\{x_0\} \times \mathbb{C}^\nu$ of the sub-D-groupoid of the Malgrange-Garnier D-groupoid, which fixes the transversals, are precisely the points of the parametrized intrinsic Galois group, that are rational over the ring $\mathbb{C}\{x - x_0\}$ of germs of analytic functions at x_0.

For systems with constant coefficients, we retrieve the result of A. Granier (*cf.* [**Gra12**, Thm. 2.4]), i.e., the evaluation in $x = x_0$ of the solutions of the transversal D-groupoid is the usual Galois group. Notice that in this case intrinsic and parametrized intrinsic Galois groups coincide. The analogous result for differential equations is proved in [**Mal01**]. B. Malgrange, in the differential case, and A. Granier, in the q-difference constant case, establish a link between the Galois

D-groupoid and the usual Galois group. This is compatible with our results since in those cases the intrinsic and parametrized intrinsic Galois groups, as well as the usual Galois groups, coincide (*cf.* §9.4 below).

Acknowledgments

We would like to thank D. Bertrand, Z. Djadli, C. Favre, M. Florence, A. Granier, D. Harari, F. Heiderich, A. Ovchinnikov, B. Malgrange, J-P. Ramis, J. Sauloy, M. Singer, J. Tapia, H. Umemura and M. Vaquie for the many discussions on different points of this paper, and the organizers of the seminars of the universities of Grenoble I, Montpellier, Rennes II, Caen, Toulouse and Bordeaux that invited us to present the results below at various stages of our work.

We would like to thank the ANR projects Diophante and qDIFF (Contract No ANR-10-JCJC 0105) and grant ECOS Nord France-Colombia No C12M01 y Colciencias "Équations aux q-différences & groupes quantiques", that have made possible a few reciprocal visits, and the Centre International de Rencontres Mathématiques in Luminy for supporting us *via* the Research in pairs program and for providing a nice working atmosphere.

Part 1

Introduction to q-difference equations

CHAPTER 1

Generalities on q-difference modules

We quickly recall some notations and a few basic results about q-difference algebra and q-difference modules. For a general introduction to difference algebra, see [**Coh65**] and [**Lev08**]. For a more detailed introduction to q-difference modules see [**vdPS97**, Chapter 12], [**DV02**, Part I] or [**DVRSZ03**].

1.1. Basic definitions

Let K be a field and $q \neq 0, 1$ be a fixed element of K. The field $K(x)$ is naturally a q-difference field, i.e., it is equipped with the q-difference operator

$$\sigma_q : \begin{array}{rcl} K(x) & \longrightarrow & K(x) \\ f(x) & \longmapsto & f(qx) \end{array}.$$

We can associate to σ_q a non-commutative derivation, that we will call q-derivation, defined by

$$d_q(f)(x) = \frac{f(qx) - f(x)}{(q-1)x},$$

and satisfying a q-Leibniz formula:

$$d_q(fg)(x) = f(qx) d_q(g)(x) + d_q(f)(x) g(x), \text{ for any } f, g \in K(x).$$

Notice that, if we set $[n]_q = \frac{q^n - 1}{q - 1}$, $[n]_q^! = [n]_q [n-1]_q \cdots [1]_q$, for any $n \geq 1$, $[0]_q^! = 1$, then

$$d_q^s x^n = \frac{[n]_q^!}{[n-s]_q^!} x^{n-s}, \text{ for any pair of positive integers } s, n, \text{ such that } n \geq s.$$

Therefore we define the q-binomial $\binom{n}{s}_q = \frac{[n]_q^!}{[n-s]_q^! [s]_q^!}$, so that $\frac{d_q^s}{[s]_q^!} x^n = \binom{n}{s}_q x^{n-s}$. When q is a root of unity of order κ, the operator d_q^κ and all its iterations are equal to 0. Nonetheless, the q-binomials $\binom{n}{s\kappa}_q$ and the operators $\frac{d_q^{s\kappa}}{[s\kappa]_q^!}$ are well-defined and non-zero for every positive integer s.

More generally, we will consider a q-difference field extension $\mathcal{F}$ of $K(x)$, i.e., a field extension $\mathcal{F}$ of $K(x)$ equipped with a field automorphism extending the action of σ_q, which we will also call q-difference operator and denote by σ_q. Of course, $\mathcal{F}$ is also equipped with the skew derivation $d_q := \frac{\sigma_q - 1}{(q-1)x}$. We denote by $\mathcal{F}^{\sigma_q}$ the field of constant of $\mathcal{F}$, i.e., the subfield of $\mathcal{F}$ of all elements fixed by σ_q.

Typical examples of q-difference field extensions of $K(x)$ are the fields $K((x))$ or $K(x^{1/r})$, for $r \in \mathbb{Z}_{>1}$. In the latter case, one sets $\sigma_q(x^{1/r}) = \tilde{q} x^{1/r}$, for a given r-th root $\tilde{q}$ of q. If $K = \mathbb{C}$, one can naturally consider also the fields of meromorphic functions over $\mathbb{C}$, over $\mathbb{C}^* = \mathbb{C} \smallsetminus \{0\}$ or over any domain invariant under the action of σ_q.

DEFINITION 1.1. A q-difference module $\mathcal{M}_{\mathcal{F}} = (M_{\mathcal{F}}, \Sigma_q)$ (of rank ν) over $\mathcal{F}$ is a finite dimensional $\mathcal{F}$-vector space $M_{\mathcal{F}}$ (of dimension ν) equipped with an invertible σ_q-semilinear operator $\Sigma_q : M_{\mathcal{F}} \to M_{\mathcal{F}}$, i.e., a bijective additive map from $M_{\mathcal{F}}$ to itself such that

$$\Sigma_q(fm) = \sigma_q(f)\Sigma_q(m), \text{ for any } f \in \mathcal{F} \text{ and } m \in M_{\mathcal{F}}.$$

We will call Σ_q a q-difference operator over $M_{\mathcal{F}}$ or the q-difference operator of $\mathcal{M}_{\mathcal{F}}$. A q-difference submodule $\mathcal{N}_{\mathcal{F}}$ of $\mathcal{M}_{\mathcal{F}}$ is an $\mathcal{F}$-vector subspace of $M_{\mathcal{F}}$ that is setwise invariant with respect to Σ_q. Then, $\mathcal{N}_{\mathcal{F}} = (N_{\mathcal{F}}, \Sigma_q|_{N_{\mathcal{F}}})$ is a q-difference module.

A morphism of q-difference modules (over $\mathcal{F}$) is a morphism of $\mathcal{F}$-vector spaces, commuting with the q-difference operators. We denote by $Diff(\mathcal{F}, \sigma_q)$ the category of q-difference modules over $\mathcal{F}$.

1.1.1. Constructions of linear algebra. Let $\mathcal{M}_{\mathcal{F}} = (M_{\mathcal{F}}, \Sigma_{q,M})$ and $\mathcal{N}_{\mathcal{F}} = (N_{\mathcal{F}}, \Sigma_{q,N})$ be two q-difference modules over $\mathcal{F}$. The direct sum $\mathcal{M}_{\mathcal{F}} \oplus \mathcal{N}_{\mathcal{F}}$ of $\mathcal{M}_{\mathcal{F}}$ and $\mathcal{N}_{\mathcal{F}}$ is the q-difference module such that:

- the underlying $\mathcal{F}$-vector space is $M_{\mathcal{F}} \oplus N_{\mathcal{F}}$;
- the q-difference operator is a σ_q-semilinear bijection defined by $m \oplus n \mapsto \Sigma_{q,M}(m) \oplus \Sigma_{q,N}(n)$.

The tensor product $\mathcal{M}_{\mathcal{F}} \otimes_{\mathcal{F}} \mathcal{N}_{\mathcal{F}}$ of $\mathcal{M}_{\mathcal{F}}$ and $\mathcal{N}_{\mathcal{F}}$ over $\mathcal{F}$ is the q-difference module such that:

- the underlying $\mathcal{F}$-vector space is $M_{\mathcal{F}} \otimes_{\mathcal{F}} N_{\mathcal{F}}$;
- the q-difference operator is a σ_q-semilinear bijection defined by $m \otimes n \mapsto \Sigma_{q,M}(m) \otimes \Sigma_{q,N}(n)$.

The dual q-difference module $\mathcal{M}_{\mathcal{F}}^* = (M_{\mathcal{F}}^*, \Sigma_{q,M}^*)$ of $\mathcal{M}_{\mathcal{F}}$ is the q-difference module defined as follows:

- the underlying $\mathcal{F}$-vector space $M_{\mathcal{F}}^*$ is the dual $\mathcal{F}$-vector space of $M_{\mathcal{F}}$;
- $\Sigma_{q,M}^* : \varphi \mapsto \sigma_q^{-1} \circ \varphi \circ \Sigma_{q,M}$, i.e., for any $m \in M_{\mathcal{F}}$ and any $\varphi \in M_{\mathcal{F}}^*$ we have $\langle \Sigma_{q,M}^*(\varphi), m \rangle = \sigma_q^{-1}\langle \varphi, \Sigma_{q,M}(m) \rangle$.

We say that a q-difference module $\mathcal{N}_{\mathcal{F}}$ over $\mathcal{F}$ is a construction of linear algebra of $\mathcal{M}_{\mathcal{F}}$ if $\mathcal{N}_{\mathcal{F}}$ can be deduced from $\mathcal{M}_{\mathcal{F}}$ by direct sums, duals, tensor products. In the Tannakian formalism, one considers also the sub-quotients of all constructions of linear algebra, but for our purpose it is enough to consider the collection of submodules of the finite direct sums of $\bigoplus \mathcal{M}_{\mathcal{F}}^{\otimes i} \otimes_{\mathcal{F}} (\mathcal{M}_{\mathcal{F}}^*)^{\otimes j}$, for any pair of non negative integers i, j (see [**And01**, §3.2.2]).

1.1.2. Basis. Let $\mathcal{M}_{\mathcal{F}} = (M_{\mathcal{F}}, \Sigma_q)$ be a q-difference module over $\mathcal{F}$ of rank ν. We fix a basis $\underline{e}$ of $M_{\mathcal{F}}$ over $\mathcal{F}$. Let $A \in \mathrm{GL}_\nu(\mathcal{F})$ be such that:

$$\Sigma_q \underline{e} = \underline{e} A.$$

If $\underline{f}$ is another basis of $M_{\mathcal{F}}$, such that $\underline{f} = \underline{e} F$, with $F \in \mathrm{GL}_\nu(\mathcal{F})$, then $\Sigma_q \underline{f} = \underline{f} B$, with $B = F^{-1} A \sigma_q(F)$.

PROPOSITION 1.2. *Let K be a field as in §1.1, $\mathcal{M}_{K(x)}$ a q-difference module over $K(x)$ and let $k = \mathbb{Q}$ or $\mathbb{F}_p$, according that the field K has characteristic zero or $p > 0$, respectively. For any q-difference module $\mathcal{M}_{K(x)}$ there exist a finitely generated extension $\widetilde{K} \subset K$ of k, containing q, and a q-difference module $\mathcal{M}_{\widetilde{K}(x)}$ such that $\mathcal{M}_{K(x)} = \mathcal{M}_{\widetilde{K}(x)} \otimes_{\widetilde{K}(x)} K(x)$.*

PROOF. To prove the lemma, it suffices to fix a basis $\underline{e}$ of $\mathcal{M}_\mathcal{F}$ and to consider a field $\widetilde{K}$ generated over k by q and all the entries of the matrix of Σ_q with respect to the basis $\underline{e}$. $\square$

REMARK 1.3. We will always denote with the same letter, but with different subscripts, q-difference modules that become isomorphic after an extension of the base field, as in the statement above.

1.1.3. Horizontal vectors. A horizontal vector of $\mathcal{M}_\mathcal{F}$ is an element $m \in \mathcal{M}_\mathcal{F}$ such that $\Sigma_q(m) = m$. We denote by $\mathcal{M}_\mathcal{F}^{\Sigma_q}$ the set of horizontal vectors of $\mathcal{M}_\mathcal{F}$. One proves easily that it is a $\mathcal{F}^{\sigma_q}$-vector space. The dimension of $\mathcal{M}_\mathcal{F}^{\Sigma_q}$ is invariant by extension of the constants:

PROPOSITION 1.4. *Let $\mathcal{F}$ be a q-difference field with $K = \mathcal{F}^{\sigma_q}$ and let K' be a field extension of K endowed with a trivial action of σ_q. Let $\mathcal{M}_\mathcal{F}$ be a q-difference module over $\mathcal{F}$ and $\mathcal{M}_{\mathcal{F}(K')} = \mathcal{M}_\mathcal{F} \otimes_\mathcal{F} \mathcal{F}(K')$ the q-difference module over $\mathcal{F}(K')$ obtained by scalar extension. Then $\left(\mathcal{M}_{\mathcal{F}(K')}\right)^{\Sigma_q} = \mathcal{M}_\mathcal{F}^{\Sigma_q} \otimes_K K'$.*

PROOF. First of all, notice that $\mathcal{F}(K')^{\sigma_q} = K'$. We have a natural injective map
$$K' \otimes_K \mathcal{M}_\mathcal{F}^{\Sigma_q} \longrightarrow \left(\mathcal{M}_{\mathcal{F}(K')}\right)^{\Sigma_q}.$$
We have to show that it is also surjective. Let $\underline{e}$ be a basis of $\mathcal{M}_\mathcal{F}$ over $\mathcal{F}$ such that $\Sigma_q \underline{e} = \underline{e} A$, with $A \in \mathrm{GL}_\nu(\mathcal{F})$. Let $z \in \left(\mathcal{M}_{\mathcal{F}(K')}\right)^{\Sigma_q}$ and let us write $z = \underline{e} Z$, where $Z \in \mathcal{F}(K')^\nu$. The set
$$\mathfrak{a} := \{r \in K' \otimes_K \mathcal{F} \text{ s.t. } rZ \in (K' \otimes_K \mathcal{F})^\nu\}$$
is a non-zero ideal of $K' \otimes_K \mathcal{F}$ stable[1] under σ_q. Indeed, if $r \in \mathfrak{a}$ then $\Sigma_q(rz) = \underline{e} A \sigma_q(rZ)$ and $A \sigma_q(rZ) \in (K' \otimes_K \mathcal{F})^\nu$. Since $\sigma_q(r)z = \Sigma_q(rz)$, we find that $\sigma_q(r) \in \mathfrak{a}$. By [**vdPS97**, Lemma 1.11], the algebra $K' \otimes_K \mathcal{F}$ has no non trivial ideal stable under σ_q. Thus 1 belongs to the ideal $\mathfrak{a}$, which implies that $Z \in (K' \otimes_K \mathcal{F})^\nu$. Let $\{\lambda_i\}_i \subset K'$ be a (maybe, infinite) basis of K'/K. We can write $z = \sum_i \lambda_i \otimes \underline{e} \vec{y}_i$, for some $\vec{y}_i \in \mathcal{F}^\nu$, not all zero. Since $\Sigma_q(z) = z$, we obtain:
$$\sum_i \lambda_i \otimes \underline{e}\vec{y}_i = \sum_i \lambda_i \otimes \underline{e} A \sigma_q(\vec{y}_i),$$
where σ_q acts on vectors componentwise. We conclude that $\vec{y}_i = A \sigma_q(\vec{y}_i)$ for all i and therefore that $\underline{e}\vec{y}_i \in \mathcal{M}_\mathcal{F}^{\Sigma_q}$, for all i. This ends the proof. $\square$

1.1.4. q-difference modules over a ring. In the sequel, we will deal with q-difference modules over rings. We do not want to be too formal on this point, since notations and definitions are quite intuitive.

Let $\mathcal{O}$ be a commutative unitary ring and $q \neq 0, 1$ be an invertible element of $\mathcal{O}$. Then σ_q defines an automorphism of the ring of polynomials $\mathcal{O}[x]$. We will call q-difference ring over $\mathcal{O}$ an $\mathcal{O}[x]$-algebra, equipped with an injective endomorphism extending σ_q. Sometimes, we shall not mention the ring $\mathcal{O}$ since the q-difference rings appearing in the next chapters will be always explicitly described, and will mainly be of the form described in the example below.

---

[1]In the whole paper, "stable " means "setwise fixed", i.e.. $\sigma_q \mathfrak{a} \subset \mathfrak{a}$.

EXAMPLE 1.5. Let $\mathcal{O}$ be a unitary subring of K containing q, q^{-1}. We will be interested in q-difference rings of the form:

$$\mathcal{O}\left[x, \frac{1}{P(x)}, \frac{1}{P(qx)}, \frac{1}{P(q^2x)}, \ldots\right],$$

for some $P(x) \in \mathcal{O}[x]$, which are subrings of $K(x)$, stable by σ_q.

Let $\mathcal{A}$ a q-difference ring over $\mathcal{O}$ and $\mathcal{A}'$ be a q-difference subring of $\mathcal{A}$ over $\mathcal{O}$, that is, an $\mathcal{O}[x]$-subalgebra of $\mathcal{A}$, stable by σ_q. We say that a q-difference ring $\mathcal{B} \subset \mathcal{A}$, containing $\mathcal{A}'$, is finitely generated (as a q-difference ring) over $\mathcal{A}'$ if there exists a finite set S such that $\mathcal{B}$ is the smallest subring of $\mathcal{A}$, containing $\mathcal{A}'$, S and stable by σ_q.

A q-difference module $\mathcal{M} = (M, \Sigma_q)$ over $\mathcal{A}$ will be a free $\mathcal{A}$-module M of finite rank, equipped with a semilinear invertible operator[2] Σ_q. All the notions introduced above generalize intuitively to this case.

If $\mathcal{A}$ is a domain and $\mathcal{F}$ is the fraction field of $\mathcal{A}$, then

$$\mathcal{M}_{\mathcal{F}} = (M_{\mathcal{F}} := M \otimes_{\mathcal{A}} \mathcal{F}, \Sigma_q \otimes \sigma_q)$$

is a q-difference module over $\mathcal{F}$.

LEMMA 1.6. *Any q-difference module over $\mathcal{F}$ comes from a q-difference module over $\mathcal{A}$, for a suitable choice of a q-difference ring $\mathcal{A} \subset \mathcal{F}$, finitely generated as q-difference ring over $\mathbb{Z}$, if the characteristic of $\mathcal{F}$ is zero, or over $\mathbb{F}_p$, otherwise.*

PROOF. Let $\mathcal{M}_{\mathcal{F}} = (M_{\mathcal{F}}, \Sigma_q)$ be a q-difference module over $\mathcal{F}$ and $k = \mathbb{Z}$ or $\mathbb{F}_p$, according to the characteristic of $\mathcal{F}$. We fix a basis $\underline{e}$ of $M_{\mathcal{F}}$ over $\mathcal{F}$ so that $\Sigma_q \underline{e} = \underline{e}A$, for some $A \in \mathrm{GL}_\nu(\mathcal{F})$. To conclude it is enough to consider the smallest ring $\mathcal{A} \subset \mathcal{F}$ containing $k[q, q^{-1}, x]$, the entries of the matrix A, the inverse of $\det A$ and stable by σ_q. □

1.2. q-difference modules, systems and equations

Let $\mathcal{M}_{\mathcal{F}} = (M_{\mathcal{F}}, \Sigma_q)$ be a q-difference module of rank ν over a q-difference field $\mathcal{F}$. We fix a basis $\underline{e}$ of $M_{\mathcal{F}}$ over $\mathcal{F}$, such that:

$$\Sigma_q \underline{e} = \underline{e}A,$$

with $A \in \mathrm{GL}_\nu(\mathcal{F})$.

DEFINITION 1.7. We call

(1.1) $$\sigma_q(Y) = A^{-1}Y,$$

the (q-difference) system (of order ν) associated to $\mathcal{M}_{\mathcal{F}}$, with respect to the basis $\underline{e}$.

If $\vec{y} \in \mathcal{F}^\nu$ are the coordinates of a horizontal vector $m \in M_{\mathcal{F}}$ with respect to the basis $\underline{e}$, then $\vec{y}$ verifies $\Sigma_q(\underline{e}\vec{y}) = \underline{e}\vec{y}$, i.e., $\vec{y} = A\sigma_q(\vec{y})$. This means that $\vec{y}$ is a solution vector of the q-difference system (1.1). On the other hand, a solution vector of (1.1) always represents a horizontal vector of $\mathcal{M}_{\mathcal{F}}$ in the basis $\underline{e}$.

---

[2]Since we will always deal with rings $\mathcal{A}$ that are domains, we could have asked that Σ_q is only injective, but then, enlarging the scalars to a q-difference algebra $\mathcal{A}'$ containing $\mathcal{A}$, constructed inverting some elements, we would have obtained an invertible operator. So for our purpose, the assumption that Σ_q is invertible is not restrictive.

Two systems are said to be equivalent by gauge transformation if they are associated to the same q-difference module, with respect to two different basis. Of course, one associates a q-difference module, with underlying $\mathcal{F}$-vector space $\mathcal{F}^\nu$, to any q-difference system of order ν.

To a given linear q-difference equation

(1.2) $\qquad a_0 y + a_1 \sigma_q y + \cdots + a_\nu \sigma_q^\nu y = 0$, with $a_1, \ldots, a_\nu \in \mathcal{F}$ and $a_0 a_\nu \neq 0$,

one naturally associates a linear q-difference system

(1.3) $\qquad \sigma_q(Y) = \begin{pmatrix} 0 & & & 1 & & 0 \\ \vdots & & & & \ddots & \\ 0 & & & 0 & & 1 \\ -a_0/a_\nu & & -a_1/a_\nu & \cdots & -a_{\nu-1}/a_\nu \end{pmatrix} Y.$

If z is a solution of (1.2) in some q-difference field extension of $\mathcal{F}$, then the vector ${}^t(z, \sigma_q(z), \ldots, \sigma_q^{\nu-1}(z))$ is a solution column of (1.3). The equation (1.2) has at most ν solutions in a q-difference field extension $\mathcal{G}$ of $\mathcal{F}$, which are linearly independent over the field $\mathcal{G}^{\sigma_q}$ of σ_q-invariant elements of $\mathcal{G}$. If $z_1, \ldots, z_\nu$ are those solutions, then the q-analog of the Wronskian Lemma see [**Cas80**, §7]) says that the matrix

$$\begin{pmatrix} z_1 & \cdots & z_\nu \\ \sigma_q(z_1) & \cdots & \sigma_q(z_\nu) \\ \vdots & \cdots & \vdots \\ \sigma_q^{\nu-1}(z_1) & \cdots & \sigma_q^{\nu-1}(z_\nu) \end{pmatrix}$$

is an invertible solution of (1.3).

Given a q-difference module $(M_\mathcal{F}, \Sigma_q)$ of rank ν over $\mathcal{F}$, such that q is not a root of unity of order smaller than ν, the Cyclic Vector Lemma (see for instance [**DV02**, §1.3]) allows to find an element m of $M_\mathcal{F}$, called cyclic element, such that $m, \Sigma_q(m), \ldots, \Sigma_q^{\nu-1}(m)$ is a basis of $M_\mathcal{F}$.

1.3. Some remarks on solutions

Let $\sigma_q(Y) = BY$ be a q-difference system, with $B \in \mathrm{GL}_\nu(\mathcal{F})$.

DEFINITION 1.8. Let $\mathcal{G}$ be a q-difference field extension of $\mathcal{F}$. A fundamental solution matrix of $\sigma_q(Y) = BY$ in $\mathcal{G}$ is an invertible matrix F, with entries in $\mathcal{G}$, such that $\sigma_q(F) = BF$.

Recursively, we obtain from $\sigma_q(Y) = BY$ a family of higher order q-difference systems:

$$\sigma_q^n(Y) = B_n Y \text{ and } d_q^n Y = G_n Y,$$

with $B_n \in \mathrm{GL}_\nu(\mathcal{F})$ and G_n in the ring $M_\nu(\mathcal{F})$ of square matrices of order ν with coefficients in $\mathcal{F}$, for any positive integer n. One can easily check that $B_1 := B$ and:

$$B_{n+1} = \sigma_q(B_n) B_1, \quad G_1 = \frac{B_1 - 1}{(q-1)x} \text{ and } G_{n+1} = \sigma_q(G_n) G_1 + d_q G_n.$$

It is convenient to set $B_0 = G_0 = 1$ and $G_{[n]} = \frac{G_n}{[n]_q^!}$ for any $n \geq 0$. Notice that $G_{[n]}$ is well-defined even if q is a root of unity.

PROPOSITION 1.9. *Let $\mathcal{F} = K(x)$ and suppose that the matrix G_1 does not have a pole at 0 (or equivalently that B does not have a pole at 0 and that $B(0)$ is the identity matrix), then $W(x) = \sum_{n\geq 0} G_{[n]}(0)x^n$ is a fundamental solution matrix (in $K((x))$) of the system $\sigma_q(Y) = BY$. Moreover, it is the only fundamental solution matrix with coefficients in $K[[x]]$, whose constant term is the identity.*

If K is a field equipped with a norm such that $|q| \neq 1$, then $\sum_{n\geq 0} G_{[n]}(0)x^n$ has a non-zero radius of convergence and, hence, an infinite radius of meromorphy.[3]

The proof of the proposition above is similar to the proof of the resolvent in the differential case. Proposition 1.9 has a multiplicative avatar:

PROPOSITION 1.10. *Let K be a field, $|\ |$ a norm (archimedean or ultrametric) over K and q an element of K, such that $|q| > 1$. We consider a q-difference system $Y(qx) = B(x)Y(x)$ such that $B(x) \in \mathrm{GL}_\nu(K(x))$, zero is not a pole of $B(x)$ and such that $B(0)$ is the identity matrix. Then the infinite product*
$$\left(B(q^{-1}x)B(q^{-2}x)B(q^{-3}x)\ldots\right)$$
is the germ at zero of the analytic fundamental solution matrix $Z(x)$ such that $Z(0)$ is the identity. Moreover, $Z(x)$ has infinite radius of meromorphy.

PROOF. If $|q| > 1$, the infinite product defining $Z(x)$ is convergent in the neighborhood of zero and it is a solution of $Y(qx) = B(x)Y(x)$, such that $Z(0)$ is the identity matrix. The fact that $Z(x)$ is a meromorphic function with infinite radius of meromorphy follows from the fact that the functional equation $Y(qx) = B(x)Y(x)$ "propagates" meromorphy. □

REMARK 1.11. Independently of the characteristic of K, if q is not a root of unity, then we can always find a norm over K such that $|q| > 1$. Of course, the norm does not need to be archimedean.

REMARK 1.12. In Proposition 1.10, if $|q| < 1$ then one has to consider the product $\prod_{n \geq 0} B(q^n x)^{-1}$.

1.4. Trivial q-difference modules

The purpose of the second part of this work is to give an arithmetic characterization of trivial q-difference modules, where trivial means:

DEFINITION 1.13. *We say that the q-difference module $\mathcal{M} = (M, \Sigma_q)$ of rank ν over a q-difference algebra $\mathcal{A}$ is trivial if there exists a basis $\underline{f}$ of M over $\mathcal{A}$ such that $\Sigma_q \underline{f} = \underline{f}$.*

The definition applies in particular to the case of a q-difference module over a field. For further reference, we state some properties of trivial q-difference modules.

PROPOSITION 1.14. *Let $\mathcal{F}$ be a q-difference field and $\mathcal{M}_\mathcal{F}$ be a q-difference module over $\mathcal{F}$. The following statements are equivalent:*
  (1) *The q-difference module $\mathcal{M}_\mathcal{F}$ is trivial.*
  (2) *There exists a basis $\underline{e}$ of $\mathcal{M}_\mathcal{F}$ such that the q-difference system associated to $\mathcal{M}_\mathcal{F}$ with respect to the basis $\underline{e}$ has an invertible solution matrix in $\mathrm{GL}_\nu(\mathcal{F})$.*

---

[3] In the sense that its entries are quotient of two entire analytic functions with respect to $|\ |$.

(3) *For any basis $\underline{e}$ of $\mathcal{M}_\mathcal{F}$, the q-difference system associated to $\mathcal{M}_\mathcal{F}$ with respect to the basis $\underline{e}$ has an invertible solution matrix in* $\mathrm{GL}_\nu(\mathcal{F})$.

(4) $\dim_{\mathcal{F}^{\sigma_q}} \mathcal{M}_\mathcal{F}^{\Sigma_q} = \dim_\mathcal{F} \mathcal{M}_\mathcal{F}$.

PROOF. Let $\underline{e}$ be a basis of $\mathcal{M}_\mathcal{F}$, such that $\Sigma_q \underline{e} = \underline{e} A(x)$, and $\underline{f}$ be a basis of $\mathcal{M}_\mathcal{F}$, such that $\underline{f} = \underline{e} F(x)$, with $F(x) \in \mathrm{GL}_\nu(\mathcal{F})$. Then $\Sigma_q \underline{f} = \underline{f}$ if and only if

$$\underline{f} = \Sigma_q(\underline{e}F(x)) = \underline{e}A(x)F(qx) = \underline{f}F(x)^{-1}A(x)F(qx),$$

therefore if and only if $F(qx) = A(x)^{-1}F(x)$. This proves the equivalence among (1), (2) and (3). The equivalence between (1) and (4) follows from the fact that $\underline{f}$ is both a basis of $M_\mathcal{F}$ over $\mathcal{F}$ and of $\mathcal{M}_\mathcal{F}^{\Sigma_q}$ over $\mathcal{F}^{\sigma_q}$. □

The following statement is a corollary of the proposition above and of Proposition 1.4:

COROLLARY 1.15. *Let K be a field, $q \neq 0, 1$ be an element of K, and $\mathcal{M}_{K(x)}$ be a q-difference module over $K(x)$. Let K' be an extension of K, on which σ_q acts as the identity, and let $\mathcal{M}_{K'(x)} = \mathcal{M}_{K(x)} \otimes_{K(x)} K'(x)$. Then $\mathcal{M}_{K(x)}$ is trivial if and only if $\mathcal{M}_{K'(x)}$ is trivial.*

PROOF. It follows from Proposition 1.4 that $\mathcal{M}_{K'(x)}^{\Sigma_q} = \mathcal{M}_{K(x)}^{\Sigma_q} \otimes_K K'$. □

Finally we consider the case of a q-difference module whose associated system has an algebraic solution over the base field $K(x)$. To the best of our knowledge, the following proposition, which nowadays is part of folklore, appears for the first time in [**Poi90**, page 318]. Here we chose a quite down-to-earth approach. For a more elegant proof see [**CS12b**, Lemme 4.4].

PROPOSITION 1.16. *Let K be a field and q be an element of K which is not a root of unity. We suppose that there exists a norm $|\ |$ over K, such that $|q| \neq 1$, and we consider a linear q-difference equation*

$$(1.4) \qquad a_\nu(x)y(q^\nu x) + a_{\nu-1}(x)y(q^{\nu-1}x) + \cdots + a_0(x)y(x) = 0$$

with coefficients in $K(x)$. If there exists an algebraic q-difference field extension $\mathcal{F}$ of $K(x)$ containing a solution f of (1.4), then f is contained in an extension of $K(x)$ isomorphic to $L(\widetilde{q}, t)$, with $\widetilde{q}^r = q, t^r = x$ and $L|K$ is a finite field extension.

PROOF. Let us look at (1.4) as an equation with coefficients in $K((x))$. Then the algebraic solution f of (1.4) can be identified to a Laurent series in $\overline{K}((t))$, where $\overline{K}$ is the algebraic closure of K and $t^r = x$, for a suitable positive integer r. Let $\widetilde{q}$ be an element of $\overline{K}$ such that $\widetilde{q}^r = q$ and that $\sigma_q(f) = f(\widetilde{q}t)$. We can look at (1.4) as a $\widetilde{q}$-difference equation with coefficients in $K(\widetilde{q}, t)$. Then the recurrence relation induced by (1.4) over the coefficients of a formal solution shows that there exist $f_1, \ldots, f_s$ solutions of (1.4) in $K(\widetilde{q})((t))$ such that $f \in \sum_i \overline{K} f_i$. It follows that there exists a finite extension $\widetilde{K}$ of $K(\widetilde{q})$ such that $f \in \widetilde{K}((t))$.

We fix an extension of $|\ |$ to $\widetilde{K}$, that we still call $|\ |$. Since f is algebraic, it is a germ of meromorphic function at 0. Since $|\widetilde{q}| \neq 1$, the functional equation (1.4) itself allows to show that f is actually a meromorphic function with infinite radius of meromorphy. Finally, if we choose r large enough, f can have at worst a pole at $t = \infty$, since it is an algebraic function, which actually implies that f is the Laurent expansion of a rational function in $\widetilde{K}(\widetilde{q}, t)$. □

We recall the following property of q-difference fields (see [**CS12a**, Lemma A.4] for the case of characteristic zero):

COROLLARY 1.17. *Let K be a field, $q \in K$ be not a root of unity and $\mathcal{M}_{K(x)}$ a q-difference module over $K(x)$. If there exists a finite q-difference field extension $\mathcal{F}$ of $K(x)$ such that $\mathcal{M}_\mathcal{F} = \mathcal{M}_{K(x)} \otimes_{K(x)} \mathcal{F}$ is trivial, then there exists a positive integer r such that $\mathcal{F} \subset L(x^{1/r})$, where $L|K$ is a finite field extension endowed with a trivial action of σ_q.*

PROOF. It is enough to apply the previous proposition to the entries of a fundamental solution matrix of the q-difference system associated to a cyclic basis of $\mathcal{M}_{K(x)}$. □

CHAPTER 2

Formal classification of singularities

2.1. Regularity

Let $\mathcal{A}$ be a q-difference subring of $K((x))$. We recall the following basic definition (see for instance [**vdPS97**] or [**Sau00**]).

DEFINITION 2.1. A q-difference module (M, Σ_q) over $\mathcal{A}$ is said to be regular singular at 0, if there exists a basis $\underline{e}$ of $(M \otimes_\mathcal{A} K((x)), \Sigma_q \otimes \sigma_q)$ over $K((x))$ such that the action of $\Sigma_q \otimes \sigma_q$ over $\underline{e}$ is represented by a constant matrix $A \in \mathrm{GL}_\nu(K)$.

We recall the following statement on regular singular q-difference modules, also known as Frobenius method or algorithm. See [**vdPS97**] or [**Sau00**, §1.1]. The Frobenius method is also briefly summarized also in [**Sau04a**, §1.2.2] and [**DVRSZ03**].

PROPOSITION 2.2. *Let $\mathcal{M}_{K((x))} = (M_{K((x))}, \Sigma_q)$ be a q-difference system over $K((x))$. We suppose that there exists a basis $\underline{e}$ of $M_{K((x))}$ such that $\Sigma_q \underline{e} = A(x)\underline{e}$, with $A(x) \in GL_\nu(K((x))) \cap M_\nu(K[[x]])$. Let $A_0 = A(0)$. Then:*
  (1) *If for any two distinct eigenvalues α and β of A_0 we have $\alpha\beta^{-1} \notin q^\mathbb{Z}$, then there exists a basis change $\underline{f} = \underline{e}F(x)$ of $M_{K((x))}$ such that $F(x) \in \mathrm{GL}_\nu(K((x))) \cap M_\nu(K[[x]])$, $F(0)$ is the identity matrix and $\Sigma_q \underline{f} = \underline{f} A_0$.*
  (2) *If the assumption of statement (1) above is not verified, a basis change in $\mathrm{GL}_\nu\left(K\left[x, \frac{1}{x}\right]\right)$, called shearing transformation, obtained multiplying alternatively invertible matrices with coefficients in K and matrices diagonal matrices whose diagonal has the form $1, \ldots, 1, x^{\pm 1}, 1, \ldots, 1$, allows to reduce to the assumption of (1).*

We point out a refinement of the statement above:

PROPOSITION 2.3 ((See [**Sau00**, §2.1.].)). *A q-difference module $M_{K(x)}$ over $K(x)$ is regular singular if and only if there exists a basis $\underline{e}$ over $K(x)$ such that $\Sigma_q \underline{e} = \underline{e} A(x)$ with $A(x) \in \mathrm{GL}_\nu(K(x)) \cap M_\nu(K[[x]])$.*

The eigenvalues of $A(0)$ are called the exponents of $\mathcal{M}$ at 0. They are well-defined modulo $q^\mathbb{Z}$. The q-difference module $\mathcal{M}$ is said to be regular singular *tout court* if it is regular singular both at 0 and at ∞, i.e., after a variable change of the form $x = 1/t$.

For further reference, we explicitly state the following lemma, which is a consequence of the Frobenius algorithm:

PROPOSITION 2.4. *Let $\mathcal{M} = (M, \Sigma_q)$ be a q-difference module over a q-difference subring $\mathcal{A}$ of $K(x)$. We assume that q is not a root of unity. The following statements are equivalent:*

(1) There exists a basis $\underline{e}$ such that $\Sigma_q \underline{e} = \underline{e}A(x)$, with $A(x) \in \mathrm{GL}_\nu(K(x)) \cap \mathrm{GL}_n(K[[x]])$, and such that $A(0)$ is a diagonal matrix with eigenvalues in $q^{\mathbb{Z}}$ (i.e., $\mathcal{M}$ has a regular singularity at 0, with integral exponents and no logarithmic singularity at 0).
(2) The q-difference module $\mathcal{M}_{K((x))}$ is trivial.

Singular regularity can be characterized with the help of a Newton polygon. Namely, regular singular q-difference modules are the ones whose Newton polygon has only one finite slope equal to 0 (see [**Sau04b**, Page 200]). We are not going to define or to list the properties of Newton polygons. We only point out that they are the key to the proof of the statements below.

Let $\mathcal{M}_{K(x)}$ be a q-difference module of rank ν and let $r \in \mathbb{N}$ be a positive integer. We consider a finite extension L of K containing an element $\widetilde{q}$ such that $\widetilde{q}^r = q$. We consider the field extension $K(x) \hookrightarrow L(t)$, $x \mapsto t^r$. The field $L(t)$ has a natural structure of $\widetilde{q}$-difference field extending the q-difference structure of $K(x)$. If follows from [**Sau04b**, §1.1.4] that:

PROPOSITION 2.5. *The q-difference module $\mathcal{M}$ is regular singular at $x = 0$ if and only if the $\widetilde{q}$-difference module $\mathcal{M}_{L(t)} := (M \otimes_{\mathcal{A}} L(t), \Sigma_{\widetilde{q}} := \Sigma_q \otimes \sigma_{\widetilde{q}})$ over $L(t)$ is regular singular at $t = 0$.*

2.2. Irregularity

Next statement gives the structure of general q-difference modules. It can be deduced from the formal classification of q-difference modules (see [**Pra83**, Corollary 9 and §9, 3)], [**Sau04b**, Theorem 3.1.6]):

PROPOSITION 2.6. *We assume that q is not a root of unity. Let $\mathcal{M}_{K(x)}$ be a q-difference module of rank ν over $K(x)$. Then there exists a positive integer r and a finite extension $L(t)$ of $K(x)$, with $t^r = x$, $r|\nu!$, and $\widetilde{q} \in L$, with $(\widetilde{q})^r = q$ such that $\mathcal{M}_{K(x)} \otimes L((t))$ is a direct sum of $\widetilde{q}$-difference modules $\mathcal{N}_i$. For any i there exists a basis $\underline{e}_i$ of $\mathcal{N}_i$ and a positive integer r_i such that $\Sigma_{\widetilde{q}} \underline{e}_i = \underline{e}_i \frac{B_i}{t^{r_i}}$, with B_i an invertible matrix with coefficients in L.*

COROLLARY 2.7. *There exist an extension $L(t)/K(x)$ as above, a basis $\underline{f}$ of the $\widetilde{q}$-difference module $\mathcal{M}_{L(t)}$ and an integer ℓ such that $\Sigma_{\widetilde{q}} \underline{f} = \underline{f} B(t)$, with $B(t) \in \mathrm{GL}_\nu(L(t))$ of the following form:*

(2.1) $$\begin{cases} B(t) = \dfrac{B_\ell}{t^\ell} + \dfrac{B_{\ell-1}}{t^{\ell-1}} + \ldots, \text{ as an element of } \mathrm{GL}_\nu(L((t))); \\ B_\ell \text{ is a constant non-nilpotent matrix.} \end{cases}$$

Part 2

Triviality of q-difference equations with rational coefficients

CHAPTER 3

Rationality of solutions, when q is an algebraic number

Let K be a field and $q \neq 0, 1$ be an element of K. We are concerned with the problem of finding a necessary and sufficient condition for a q-difference module $\mathcal{M}_{K(x)} = (M_{K(x)}, \Sigma_q)$ over $K(x)$ to be trivial (see Definition 1.13). This is equivalent to the problem of finding a necessary and sufficient condition for a linear q-difference system with coefficients in $K(x)$ to have a fundamental solution matrix with entries in $K(x)$.

Notice that we are not making any assumption on the characteristic of K. We have to consider the following cases:
  (1) q is a root of unity;
  (2) q is algebraic over the prime field, but is not a root of unity;
  (3) q is transcendental over the prime field.

These six cases (three cases for the characteristic zero, and three cases for the positive one) actually boil down to three. In fact, we will first consider the (trivial) situation in which q is a root of unity: If K has positive characteristic this includes both (1) and (2) above. Then we will consider the case in which K has characteristic zero and q is algebraic over $\mathbb{Q}$. Finally, in the next chapter, we will consider the case in which q is transcendental over the prime field, $\mathbb{Q}$ or $\mathbb{F}_p$, regardless of the characteristic.

It is not difficult to prove that:

PROPOSITION 3.1 ([**Hen96**] or [**DV02**, Proposition 2.1.2]). *If q is a primitive root of unity of order κ, a q-difference module $\mathcal{M}_{K(x)}$ over $K(x)$ is trivial if and only if Σ_q^κ is the identity.*

The proposition above completes the study of the triviality of q-difference modules when q is a root of unity, at least as far as the problem we are considering here is regarded. We refer to [**Har10**] for a more sophisticated approach.

3.1. The case of q algebraic, not a root of unity

If q is algebraic, but not a root of unity, we are necessarily in characteristic zero. The example below gives the guidelines for the whole chapter.

EXAMPLE 3.2. Let $K = \mathbb{Q}(a)$ be a purely transcendental extension of degree 1 and let $q \in \mathbb{Q} \smallsetminus \{0, 1, -1\}$. We consider a q-difference module $\mathcal{M}_{K(x)} = (M_{K(x)}, \Sigma_q)$ over $K(x)$. Let us choose a basis $\underline{e}$ of $M_{K(x)}$ and let $Y(qx) = B(a, x)Y(x)$ be the associated q-difference system. One can construct a $\mathbb{Z}$-algebra stable under σ_q, of the form:
$$\mathcal{A} = \mathbb{Z}\left[a, x, \frac{1}{P(x)}, \frac{1}{P(qx)}, \dots\right],$$

for a suitable choice of $P(x) \in \mathbb{Z}[\underline{a},x]$, such that $q \in \mathcal{A}$ and $B(a,x)$ and $B(a,x)^{-1}$ are both matrices with coefficients in $\mathcal{A}$. For almost all primes p in $\mathbb{Z}$, one can reduce both q and $\mathcal{A}$ modulo p, and hence the coefficients of $B(a,x)$. In particular, for all such p's, there exist a minimal positive integer κ_p and a positive integer ℓ_p, such that $q^{\kappa_p} \equiv 1$ modulo p and $q^{\kappa_p} - 1 = p^{\ell_p} \frac{r}{s}$, with r,s prime to p. The main result of this chapter (see Theorem 3.6 below) is that the system $Y(qx) = B(a,x)Y(x)$ has a fundamental solution matrix with coefficients in $K(x)$ if and only if for almost all p we have

$$(3.1) \qquad B(a, q^{\kappa_p-1}x)B(a, q^{\kappa_p-2}x) \cdots B(a,x) \equiv 1 \text{ modulo } p^{\ell_p}, \text{ i.e., in } \mathcal{A}/p^{\ell_p}\mathcal{A}.$$

This last condition is equivalent to the fact that the reduction modulo p^{ℓ_p} of the operator $\Sigma_q^{\kappa_p}$ is the identity, and is verified, in particular, if the reduction of the system $Y(qx) = B(a,x)Y(x)$ modulo p^{ℓ_p} has a fundamental solution matrix with coefficients in $\mathcal{A}/p^{\ell_p}\mathcal{A}$. We will proceed as follows: We will first prove that the system $Y(qx) = B(a,x)Y(x)$ has a fundamental solution matrix with coefficients in $K(x)$ if and only if, for all α in a dense subset of the algebraic closure $\overline{\mathbb{Q}}$ of $\mathbb{Q}$, the system $Y(qx) = B(\alpha,x)Y(x)$ has a fundamental solution matrix with coefficients in $\overline{\mathbb{Q}}(x)$. As a consequence of [**DV02**, Theorem 7.1.1], we will show that this last condition, holding for all α in a dense subset of $\overline{\mathbb{Q}}$, is equivalent to (3.1).

First of all we need to introduce some notation, that generalizes the one in the previous example to the case of a number field. With no loss of generality, we will assume that K is finitely generated over $\mathbb{Q}$ (see Proposition 1.2). Let Q be the algebraic closure of $\mathbb{Q}$ inside K. Then the field K has the form $Q(\underline{a},b)$, where $\underline{a} = (a_1, \ldots, a_r)$ is a transcendence basis of K/Q and b is a primitive element of the algebraic extension $K/Q(\underline{a})$. We call $\mathcal{O}_Q$ the ring of integers of Q, v a finite place of Q and π_v a v-adic uniformizer in $\mathcal{O}_Q$.

We fix an element $q \in K$ which is algebraic over $\mathbb{Q}$ and not a root of unity, i.e., an element $q \in Q$ which is not a root of unity. For almost all v,

- the order κ_v of q modulo v, as a root of unity,
- the positive integer power ϕ_v of π_v, such that $\phi_v^{-1}(1 - q^{\kappa_v})$ is a unit of $\mathcal{O}_Q$,

are well-defined.

We consider a q-difference module $\mathcal{M}_{K(x)} = (M_{K(x)}, \Sigma_q)$ over $K(x)$, of finite rank ν. Choosing conveniently the set of generators $\underline{a}, b$ of K/Q, we can always find a q-difference algebra $\mathcal{A}$ of the form:

$$(3.2) \qquad \mathcal{A} = \mathcal{O}_Q\left[\underline{a}, b, x, \frac{1}{P(x)}, \frac{1}{P(qx)}, \ldots\right],$$

for some $P(x) \in \mathcal{O}_Q[\underline{a}, b, x]$, and a Σ_q-stable $\mathcal{A}$-lattice M of $\mathcal{M}_{K(x)}$ such that the restriction of Σ_q to M is invertible. According to the definition in §1.1.4, the pair $\mathcal{M} = (M, \Sigma_q)$ is a q-difference module over the ring $\mathcal{A}$.

NOTATION 3.3. For a given q-difference module $\mathcal{M}_{K(x)} = (M_{K(x)}, \Sigma_q)$ over $K(x)$, the pair $\mathcal{M} = (M, \Sigma_q)$ will always denote a q-difference module over a ring $\mathcal{A}$ as above, such that $\mathcal{M} \otimes_{\mathcal{A}} K(x) := (M \otimes_{\mathcal{A}} K(x), \Sigma_q \otimes_{\mathcal{A}} \sigma_q) \cong \mathcal{M}_{K(x)}$. The notation may appear ambiguous, but it is actually convenient and there will be no confusion. We will come back on the implications of the choice of $\mathcal{A}$ and $\mathcal{M}$ in Remark 3.5 below.

3.1. THE CASE OF q ALGEBRAIC, NOT A ROOT OF UNITY

DEFINITION 3.4. We say that a q-difference module $\mathcal{M} = (M, \Sigma_q)$ over a q-difference $\mathcal{O}_Q$-algebra $\mathcal{A}$, as above, has zero v-curvature modulo ϕ_v if the linear operator
$$\Sigma_q^{\kappa_v} : M \otimes_\mathcal{A} \mathcal{A}/(\phi_v) \longrightarrow M \otimes_\mathcal{A} \mathcal{A}/(\phi_v)$$
is the identity. By abuse of language we will say that the q-difference module $\mathcal{M}_{K(x)} = \mathcal{M} \otimes_\mathcal{A} K(x)$ has zero v-curvature modulo ϕ_v, if $\mathcal{M}$ does.

REMARK 3.5. (1) First of all, the definition is justified by the fact that $\Sigma_q^{\kappa_v}$ induces the identity modulo ϕ_v if and only if $(\Delta_q)^{\kappa_v}$, where $\Delta_q = \frac{\Sigma_q - 1}{(q-1)x}$, is zero modulo ϕ_v. Therefore the terminology is inspired by the classical terminology for differential equations, [**Kat70**]. Exactly as it happens in p-adic theory of differential equations, the property of having v-curvature zero can be reformulated in a more analytic way, in terms of v-adic radius of convergence of solutions of a q-difference system associated to $\mathcal{M}$. This point of view plays no direct role in the present paper. For more details, see [**DV02**, Part II], and more precisely Corollary 5.2.2.
(2) Secondly, we point out that the quotient $\mathcal{O}_Q/(\phi_v)$ is not an integral domain in general. Nonetheless the following implication is always true. If $M \otimes_\mathcal{A} \mathcal{A}/(\phi_v)$, equipped with the operator induced by Σ_q, is trivial as a q-difference module over $\mathcal{A}/(\phi_v)$, then $\Sigma_q^{\kappa_v}$ induces the identity modulo ϕ_v. The converse is not true in such generality (see [**DV02**, Proposition 2.1.2]), unless we make either the local assumption that $\Sigma_q^{\kappa_v}$ induces the identity modulo π_v or the global assumption that $\Sigma_q^{\kappa_v}$ induces the identity modulo ϕ_v for almost all v. See the theorem below.
(3) Notice that the reduction modulo ϕ_v of $\Sigma_q^{\kappa_v}$ is well-defined, for almost all finite places v of Q. Moreover, given two q-difference modules $\mathcal{M}$ over $\mathcal{A}$ and $\mathcal{M}'$ over $\mathcal{A}'$, such that $\mathcal{M} \otimes_\mathcal{A} K(x) \cong \mathcal{M}' \otimes_{\mathcal{A}'} K(x)$, the reduction modulo ϕ_v of the first one has zero v-curvature if and only if also the other does, provided that ϕ_v is not invertible in both $\mathcal{A}$ and $\mathcal{A}'$.

Our first result is the following:

THEOREM 3.6. *A q-difference module $\mathcal{M}$ over $\mathcal{A}$ has zero v-curvature modulo ϕ_v, for almost all finite places v of Q, if and only if $\mathcal{M}_{K(x)}$ is trivial.*

REMARK 3.7. The theorem above is proved in [**DV02**] under the assumption that K is a number field, i.e., that $Q = K$. Here K is only a finitely generated extension of $\mathbb{Q}$. Notice the proof below relies crucially on [**DV02**], but is not a generalization of the arguments in [**DV02**].

If the q-difference module $\mathcal{M}_{K(x)}$ over $K(x)$ is trivial, it is not difficult to show that $\mathcal{M}$ has zero v-curvature modulo ϕ_v, for almost all finite places v of Q, for any choice of $\mathcal{A}$ and $\mathcal{M}$, such that $\mathcal{M} \otimes_\mathcal{A} K(x) \cong \mathcal{M}_{K(x)}$. So we only have to prove the inverse implication.

We are actually going to prove a stronger result:

THEOREM 3.8. *A q-difference module $\mathcal{M}$ over $\mathcal{A}$ has zero v-curvature modulo ϕ_v, for all places v in a set S of finite places of Q of Dirichlet density 1 if and only if $\mathcal{M}_{K(x)}$ is trivial.*

We recall that a subset S of the set of finite places $\mathcal{C}$ of Q has Dirichlet density 1 if

$$(3.3) \qquad \limsup_{s \to 1^+} \frac{\sum_{v \in S, v|p} p^{-sf_v}}{\sum_{v \in \mathcal{C}, v|p} p^{-sf_v}} = 1,$$

where f_v is the degree of the residue field of v over $\mathbb{F}_p$.

3.2. Global nilpotence.

We start proving a result of regularity (see §2.1 for the definition), inspired by [**Kat70**].

DEFINITION 3.9. We say that a q-difference module $\mathcal{M} = (M, \Sigma_q)$ over a q-difference $\mathcal{O}_Q$-algebra $\mathcal{A}$, as above, has nilpotent v-curvature modulo π_v, or simply that it has nilpotent reduction modulo π_v, if the linear operator

$$\Sigma_q^{\kappa_v} : M \otimes_\mathcal{A} \mathcal{A}/(\pi_v) \longrightarrow M \otimes_\mathcal{A} \mathcal{A}/(\pi_v)$$

is unipotent (or equivalently, if the linear operator induced by $\Delta_q^{\kappa_v}$ is nilpotent. See [**DV02**, §2]).

PROPOSITION 3.10. *Let $\mathcal{M} = (M, \Sigma_q)$ be a q-difference module over a q-difference $\mathcal{O}_Q$-algebra $\mathcal{A}$ of the form (3.2).*
  (1) *If $\mathcal{M}$ has nilpotent v-curvature modulo π_v, for infinitely many finite places v of Q, then the q-difference module $\mathcal{M}_{K(x)}$ is regular singular.*
  (2) *If there exists a set S of finite places v of Q of Dirichlet density 1 such that $\mathcal{M}$ has nilpotent v-curvature modulo π_v, for all $v \in S$, then $\mathcal{M}_{K((x))}$ is trivial.*

The proof of Proposition 3.10 is almost the same as [**DV02**, Theorem 6.2.2 and Proposition 6.2.3]. The last sentence of the proof of 1) in *loc. cit.* needs to be rectified, so that we prefer to repeat the proof here. We recall the following key-proposition:

PROPOSITION 3.11 ([**DV02**, Proposition 6.1.1]). *Let S be a set of finite places of Q of Dirichlet density equal to 1. If a and b are two non-zero elements of Q, not roots of unity, such that*
(1) for all $v \in S$, the reduction of a et b modulo π_v is well-defined and non-zero;
(2) for all $v \in S$, the reduction modulo π_v of b belongs to the cyclic group generated by the reduction modulo π_v of a.
Then $b \in a^{\mathbb{Z}}$.

PROOF OF PROPOSITION 3.10. To prove assertion (1), it is enough to prove that 0 is a regular singular point for $\mathcal{M}$, the proof at ∞ being completely analogous.

In the notation of Corollary 2.7, we consider the extension $L(t)$ of $K(x)$, the $\tilde{q}$-difference module $\mathcal{M}_{L(t)}$ obtained by scalar extension and the basis $\underline{f}$ such that $\Sigma_{\tilde{q}}\underline{f} = \underline{f}B(t)$, with $B(t)$ as in (2.1). Let $\tilde{Q}$ be the algebraic closure of $\mathbb{Q}$ in L and $\mathcal{B} \subset L(t)$ be a $\tilde{q}$-difference algebra over the ring of integers $\mathcal{O}_{\tilde{Q}}$ of $\tilde{Q}$, of the same form as (3.2), containing the entries of $B(t)$ and the inverse of its determinant. Let w be a finite place of $\tilde{Q}$ and $\pi_w \in \tilde{Q}$ be the uniformizer of w. Then there exists a $\tilde{q}$-difference module $\mathcal{N}$ over $\mathcal{B}$ such that $\mathcal{N} \otimes_\mathcal{B} L(t) \cong \mathcal{M}_{L(t)}$, having the following properties:

1. $\mathcal{N}$ has nilpotent w-curvature modulo π_w, for infinitely many finite places w of $\widetilde{Q}$;
2. there exists a basis $\underline{f}$ of $\mathcal{N}$ over $\mathcal{B}$ such that $\Sigma_{\widetilde{q}}\underline{f} = \underline{f}B(t)$ and $B(t)$ verifies Corollary 2.7 and in particular (2.1), i.e., it can be written in the form $B(t) = \frac{B_\ell}{t^\ell} + \frac{B_{\ell-1}}{t^{\ell-1}} + \cdots \in \mathrm{GL}_\nu(L((t)))$ for some $\ell \in \mathbb{Z}$ and B_ℓ is a constant non-nilpotent matrix, with coefficients in L.

Iterating the operator $\Sigma_{\widetilde{q}}$ we obtain:

$$\Sigma_{\widetilde{q}}^m(\underline{f}) = \underline{f}B(t)B(\widetilde{q}t)\cdots B(\widetilde{q}^{m-1}t) = \underline{f}\left(\frac{B_\ell^m}{\widetilde{q}^{\frac{\ell m(\ell m-1)}{2}}t^{m\ell}} + h.o.t.\right).$$

We know that, for infinitely many finite places w of $\widetilde{Q}$, the matrix $B(t)$ verifies

$$(3.4) \qquad \left(B(t)B(\widetilde{q}t)\cdots B(\widetilde{q}^{\kappa_w-1}t) - 1\right)^{n(w)} \equiv 0 \bmod \pi_w,$$

where κ_w is the order $\widetilde{q}$ modulo π_w and $n(w)$ is a suitable positive integer. Suppose that $\ell \neq 0$. Then $B_\ell^{\kappa_w n(w)} \equiv 0$ modulo π_w, for infinitely many w, and hence B_ℓ is a nilpotent matrix, in contradiction with Corollary 2.7. So necessarily $\ell = 0$.

Finally we have $\Sigma_{\widetilde{q}}(\underline{f}) = \underline{f}(B_0 + h.o.t)$. It follows from (3.4) that B_0 is actually invertible, which implies that $\mathcal{M}_{L(t)}$ is regular singular at 0. Proposition 2.5 allows to end the proof of (1).

Let us prove the second part of Proposition 3.10. We have already proved that 0 is a regular singularity for $\mathcal{M}$. This means that there exists a basis $\underline{e}$ of $\mathcal{M}_{K(x)}$ over $K(x)$ such that $\Sigma_q \underline{e} = \underline{e}A(x)$, with $A(x) \in \mathrm{GL}_\nu(K[[x]]) \cap \mathrm{GL}_\nu(K(x))$.

The Frobenius algorithm (See Proposition 2.2 and [**Sau00**, §1.1.1]) implies that there exists a shearing transformation $S \in \mathrm{GL}_\nu(K[x, 1/x])$, such that $S(qx)A(x)S(x)^{-1} \in \mathrm{GL}_\nu(K[[x]]) \cap \mathrm{GL}_\nu(K(x))$ and that the constant term A_0 of $S(x)^{-1}A(x)S(qx)$ has the following properties: if α and β are eigenvalues of A_0 and $\alpha\beta^{-1} \in q^{\mathbb{Z}}$, then $\alpha = \beta$. So choosing the basis $\underline{e}S(x)$ instead of $\underline{e}$, we can assume that $A_0 = A(0)$ has this last property.

Always following the Frobenius algorithm, one constructs recursively the entries of a matrix $F(x) \in \mathrm{GL}_\nu(K[[x]]))$, with $F(0) = 1$, such that we have $F(x)^{-1}A(x)F(qx) = A_0$. This means that there exists a basis $\underline{f}$ of $\mathcal{M}_{K((x))}$ such that $\Sigma_q \underline{f} = \underline{f}A_0$.

The matrix A_0 can be written as the product of a semi-simple matrix and a unipotent matrix. Since $\mathcal{M}$ has nilpotent reduction modulo π_v, we deduce that the reduction of $A_0^{\kappa_v}$ modulo π_v is the identity matrix, for any $v \in S$. First of all, this implies that A_0 is diagonalisable. Let $\widetilde{K}$ be a finite extension of K in which we can find all the eigenvalues of A_0. Then any eigenvalue $\alpha \in \widetilde{K}$ of A_0 has the property that $\alpha^{\kappa_v} = 1$ modulo π_w, for all w finite place of the algebraic closure of Q in $\widetilde{K}$ such that $w|v$ and $v \in S$. In other words, the reduction modulo w of an eigenvalue α of A_0 belongs to the multiplicative cyclic group generated by the reduction of q modulo the uniformizer π_w of w. Proposition 3.11 implies that $\alpha \in q^{\mathbb{Z}}$. We conclude appling Proposition 2.4. $\square$

3.3. Proof of Theorem 3.8 (and of Theorem 3.6)

Notice that Theorem 3.6 is a special case of Theorem 3.8. The proof of Theorem 3.8 is divided into steps. We remind that, if K is finite over $\mathbb{Q}$, the statement is proved in [**DV02**].

STEP 0. REDUCTION TO A PURELY TRANSCENDENTAL EXTENSION K/Q.
Let $\underline{a}$ be a transcendence basis of K/Q and b is a primitive element of $K/Q(\underline{a})$, so that $K = Q(\underline{a}, b)$. By restriction of scalars, the module $\mathcal{M}_{K(x)}$ is also a q-difference module of finite rank over $Q(\underline{a})(x)$. Since the field $K(x)$ is a trivial q-difference module over $Q(\underline{a})(x)$, we have:

- the module $\mathcal{M}_{K(x)}$ is trivial over $K(x)$ if and only if it is trivial over $Q(\underline{a})(x)$ (see Corollary 1.15);
- under the present assumptions, there exist an algebra $\mathcal{A}'$ of the form

$$(3.5) \quad \mathcal{A}' = \mathcal{O}_Q\left[\underline{a}, x, \frac{1}{R(x)}, \frac{1}{R(qx)}, \ldots\right], \text{ with } R(x) \in \mathcal{O}_Q[\underline{a}, x],$$

and a $\mathcal{A}'$-lattice $\mathcal{M}_{\mathcal{A}'}$ of q-difference module $\mathcal{M}_{K(x)}$ over $Q(\underline{a})(x)$, such that $\mathcal{M}_{\mathcal{A}'} \otimes_{\mathcal{A}'} Q(\underline{a}, x) = \mathcal{M}_{K(x)}$, as a q-difference module over $Q(\underline{a}, x)$, and $\Sigma_q^{\kappa_v}$ induces the identity on $\mathcal{M}_{\mathcal{A}'} \otimes_{\mathcal{A}'} \mathcal{A}'/(\phi_v)$, for all places $v \in S$.

For this reason, we can actually assume that K is a purely transcendental extension of Q of degree $d > 0$ and that $\mathcal{A} = \mathcal{A}'$. We fix an immersion of $Q \hookrightarrow \overline{Q}$, so that we will think to the transcendental basis $\underline{a}$ as a set of parameter generically varying in $\overline{Q}^d$. □

STEP 0BIS. INITIAL DATA. Let $K = Q(\underline{a})$ and q be a non-zero element of Q, which is not a root of unity. We are given a q-difference module $\mathcal{M}$ over a suitable algebra $\mathcal{A}$ as above, such that $K(x)$ is the field of fraction of $\mathcal{A}$ and such that $\Sigma_q^{\kappa_v}$ induces the identity on $M \otimes_{\mathcal{A}} \mathcal{A}/(\phi_v)$, for all finite places $v \in S$. We fix a basis $\underline{e}$ of $\mathcal{M}$, such that $\Sigma_q \underline{e} = \underline{e} A^{-1}(x)$, with $A(x) \in \mathrm{GL}_\nu(\mathcal{A})$. We will rather work with the associated q-difference system:

$$(3.6) \quad Y(qx) = A(x)Y(x).$$

It follows from Proposition 3.10 that $\mathcal{M}_{K(x)}$ is regular singular, with no logarithmic singularities, and that its exponents are in $q^{\mathbb{Z}}$ (see also Proposition 2.4). Enlarging a little bit the algebra $\mathcal{A}$ (more precisely replacing the polynomial R by a multiple of R), we can suppose that both 0 and ∞ are not poles of $A(x)$ and that $A(0), A(\infty)$ are *simultaneously* diagonal matrices with eigenvalues in $q^{\mathbb{Z}}$ (see [**Sau00**, Theoreme §2.1]). □

STEP 1. CONSTRUCTION OF A FUNDAMENTAL SOLUTION MATRIX AT 0. We construct a fundamental matrix of solutions, applying the Frobenius algorithm to this particular situation. See Proposition 2.2. There exists a shearing transformation $S_0(x) \in \mathrm{GL}_\nu(K[x, x^{-1}])$ such that

$$S_0^{-1}(qx)A(x)S_0(x) = A_0(x)$$

and $A_0(0)$ is the identity matrix. Since $A(0)$ is a diagonal matrix with eigenvalues in $q^{\mathbb{Z}}$, the matrix $S_0(x)$ can be written as a product of diagonal matrices with integer powers of x on the diagonal and therefore belongs to $\mathrm{GL}_\nu(Q[x, x^{-1}])$ (see [**Sau00**, Example 1 and 2 in §1.1.1]). Once again, up to a finitely generated extension of the algebra $\mathcal{A}$, obtained inverting a suitable polynomial, we can suppose that $S_0(x) \in \mathrm{GL}_\nu(\mathcal{A})$.

Notice that, since q is not a root of unity, there always exists a norm, non-necessarily archimedean, on Q such that $|q| > 1$. We can always extend such a norm to K, giving an arbitrary value to the elements of a basis of transcendence (see [**Bou64**, §2.4]). As in Proposition 1.10, the system

(3.7) $$Z(qx) = A_0(x)Z(x)$$

has a unique convergent solution $Z_0(x)$, such that $Z_0(0)$ is the identity and $Z_0(x)$ is a germ of a meromorphic function with infinite radius of meromorphy. So we have the following meromorphic solution of $Y(qx) = A(x)Y(x)$:

$$Y_0(x) = \left(A_0(q^{-1}x)A_0(q^{-2}x)A_0(q^{-3}x)\ldots\right)S_0(x).$$

We remind that this infinite product represents a meromorphic fundamental solution matrix of $Y(qx) = A(x)Y(x)$ for any norm over K such that $|q| > 1$. □

STEP 2. CONSTRUCTION OF A FUNDAMENTAL SOLUTION MATRIX AT ∞. In exactly the same way, we can construct a solution at ∞ of the form $Y_\infty(x) = Z_\infty(x)S_\infty(x)$, where the matrix S_∞ belongs to $GL_\nu(K[x, x^{-1}]) \cap \mathrm{GL}_\nu(\mathcal{A})$ and has the same form as $S_0(x)$, and $Z_\infty(x)$ is analytic in a neighborhood of ∞, with $Z_\infty(\infty) = 1$: $Y_\infty(x) = \left(A_\infty(x)A_\infty(qx)A_\infty(q^2x)\ldots\right)S_\infty(x)$. □

STEP 3. THE BIRKHOFF MATRIX. To summarize we have constructed two fundamental solution matrices, $Y_0(x)$ at zero and $Y_\infty(x)$ at ∞, which are meromorphic over $\mathbb{A}^1_K \smallsetminus \{0\}$, for any norm on K such that $|q| > 1$, and such that their set of non-zero poles and zeros is contained in the q-orbits of the set of poles at zeros of $A(x)$ and $A(x)^{-1}$. The Birkhoff matrix

$$B(x) = Y_0^{-1}(x)Y_\infty(x) = S_0(x)^{-1}Z_0(x)^{-1}Z_\infty(x)S_\infty(x)$$

is a meromorphic matrix on $\mathbb{A}^1_K \smallsetminus \{0\}$ with elliptic entries, i.e., $B(qx) = B(x)$. All the zeros and poles of $B(x)$, other than 0 and ∞, are contained in the q-orbits of the zeros and poles of the matrices $A(x)$ and $A(x)^{-1}$ (see [**Sau00**, §2.3.1]). □

STEP 4. RATIONALITY OF THE BIRKHOFF MATRIX. Let us choose $\underline{\alpha} = (\alpha_1, \ldots, \alpha_r)$, with α_i in the algebraic closure $\overline{\mathbb{Q}}$ of $\mathbb{Q}$, so that we can specialize $\underline{a}$ to $\underline{\alpha}$ in the coefficients of $A(x), A(x)^{-1}, S_0(x), S_\infty(x)$ and that the specialized matrices are still invertible. Then we obtain a q-difference system with coefficients in $\mathbb{Q}(\underline{\alpha})$. It follows from Proposition 1.10 that for any norm on $\mathbb{Q}(\underline{\alpha})$ such that $|q| > 1$, we can specialize $Y_0(x), Y_\infty(x)$ and, therefore $B(x)$, to matrices with meromorphic entries on $\mathbb{Q}(\underline{\alpha})^*$. We will write $A^{(\underline{\alpha})}(x), Y_0^{(\underline{\alpha})}(x)$, etc. for the specialized matrices. Notice that, since $S_0, S_\infty \in \mathrm{GL}_\nu(\mathbb{Q}(x))$, the matrix $Y_0^{(\underline{\alpha})}(x)$ is such that $S_0^{-1}(x)Y_0^{(\underline{\alpha})}(x)$ is the only fundamental solution matrix of $Z(qx) = A_0^{(\underline{\alpha})}(x)Z(x)$, analytic at zero, whose constant term is the identity (see Proposition 1.9). An analogous statement holds for $Y_\infty^{(\underline{\alpha})}(x)$.

For almost all v, it still makes sense to reduce $A_{\kappa_v}^{(\underline{\alpha})}(x)$ modulo ϕ_v. Moreover, since $A_{\kappa_v}(x)$ is the identity modulo ϕ_v, the same holds for $A_{\kappa_v}^{(\underline{\alpha})}(x)$. Therefore the reduced system has zero v-curvature modulo ϕ_v, for almost all $v \in S$. We know from [**DV02**, Theorem 7.1.1], that $Y_0^{(\underline{\alpha})}(x)$ and $Y_\infty^{(\underline{\alpha})}(x)$ are, respectively, the germs at zero and at ∞ of rational functions, and therefore that $B^{(\underline{\alpha})}(x)$ is a constant matrix in $\mathrm{GL}_\nu(\mathbb{Q}(\underline{\alpha}))$.

As we have already pointed out, $B(x)$ is q-invariant meromorphic matrix on $\mathbb{P}^1_K \smallsetminus \{0, \infty\}$. The set of its poles and zeros is the union of a finite numbers of q-orbits of the forms $\beta q^{\mathbb{Z}}$, such that β is algebraic over K and is a pole or a zero of $A(x)$ or

$A(x)^{-1}$. If β is a pole or a zero of an entry $b(x)$ of $B(x)$ and $h_\beta(x), k_\beta(x) \in Q[\underline{a}, x]$ are the minimal polynomials of β and β^{-1} over K, respectively, then we have:

$$b(x) = \lambda \frac{\prod_\gamma \prod_{n\geq 0} h_\gamma(q^{-n}x) \prod_{n\geq 0} k_\gamma(1/q^n x)}{\prod_\delta \prod_{n\geq 0} h_\delta(q^{-n}x) \prod_{n\geq 0} k_\delta(1/q^n x)},$$

where $\lambda \in K$ and γ and δ vary in a system of representatives of the q-orbits of the zeroes and the poles of $b(x)$, respectively. We have proved that there exists a dense subset of $\overline{\mathbb{Q}}^d$ such that the specialization of $b(x)$ at any point of this set is constant. Since the factorization written above must specialize to a convergent factorization of the same form of the corresponding element of $B^{(\underline{\alpha})}(x)$, we conclude that $b(x)$, and therefore $B(x)$, is a constant. $\square$

The fact that $B(x) \in \mathrm{GL}_\nu(K)$ implies that the solutions $Y_0(x)$ and $Y_\infty(x)$ glue to a meromorphic solution on $\mathbb{P}^1_K$ and ends the proof of Theorem 3.6.

CHAPTER 4

Rationality of solutions when q is transcendental

In this chapter we consider the case of q transcendental over the prime field.

4.1. Statement of the main result

Let us consider the field of rational function $k(q)$ with coefficients in a perfect field k, of any characteristic.[1] We fix $d \in]0,1[$ and for any irreducible polynomial $v = v(q) \in k[q]$ we set:

$$|f(q)|_v = d^{\deg_q v(q) \cdot \mathrm{ord}_{v(q)} f(q)}, \quad \forall f(q) \in k[q].$$

The definition of $|\ |_v$ extends to $k(q)$ by multiplicativity. To this set of norms one has to add the q^{-1}-adic one, defined on $k[q]$ by:

$$|f(q)|_{q^{-1}} = d^{-\deg_q f(q)}.$$

Once again, this definition extends by multiplicativity to $k(q)$. Then, the product formula holds:

$$\prod_{v \in k[q] \text{ irred.}} \left|\frac{f(q)}{g(q)}\right|_v = d^{\sum_v \deg_q v(q) \left(\mathrm{ord}_{v(q)} f(q) - \mathrm{ord}_{v(q)} g(q)\right)}$$
$$= d^{\deg_q f(q) - \deg_q g(q)}$$
$$= \left|\frac{f(q)}{g(q)}\right|_{q^{-1}}^{-1}.$$

For any finite extension K of $k(q)$, we consider the family $\mathcal{P}$ of ultrametric norms, that extends the norms defined above, up to equivalence. We suppose that the norms in $\mathcal{P}$ are normalized so that the product formula still holds. We consider the following partition of $\mathcal{P}$:

- the set $\mathcal{P}_\infty$ of places of K such that the associated norms extend, up to equivalence, either $|\ |_q$ or $|\ |_{q^{-1}}$;
- the set $\mathcal{P}_f$ of places of K such that the associated norms extend, up to equivalence, one of the norms $|\ |_v$ for an irreducible $v = v(q) \in k[q]$, $v(q) \neq q$.[2]

---

[1] One can always replace a field in positive characteristic with its perfect closure. See Theorem 6.14 below.

[2] The notation $\mathcal{P}_f$, $\mathcal{P}_\infty$ is only psychological, since all the norms involved here are ultrametric. Nevertheless, there exists a fundamental difference between the two sets, in fact for any $v \in \mathcal{P}_\infty$ one has $|q|_v \neq 1$, while for any $v \in \mathcal{P}_f$ the v-adic norm of q is 1. Therefore, from a v-adic analytic point of view, a q-difference equation has a totally different nature with respect to the norms in the sets $\mathcal{P}_f$ or $\mathcal{P}_\infty$. The dichotomy between $\mathcal{P}_f$ and $\mathcal{P}_\infty$ can also be explained through the geometry of q-difference equations. Indeed the action of σ_q over $K(x)$ can be encoded in an action of $\mathbb{G}_m$ over $\mathbb{P}^1_K$, whose closed orbits are the places in $\mathcal{P}_\infty$, while the open orbits are the places in $\mathcal{P}_f$.

Moreover we consider the set $\mathcal{C}$ of places $v \in \mathcal{P}_f$ such that v divides a valuation of $k(q)$ having as uniformizer a factor of a cyclotomic polynomial, other than $q-1$. Equivalently, $\mathcal{C}$ is the set of places $v \in \mathcal{P}_f$ such that q reduces to a root of unity modulo v of order κ_v strictly greater than 1. We call $v \in \mathcal{C}$ a cyclotomic place.

Sometimes we will write $\mathcal{P}_K, \mathcal{P}_{K,f}, \mathcal{P}_{K,\infty}$ and $\mathcal{C}_K$, to stress out the choice of the base field.

In the sequel, we will deal with an arithmetic situation, in the following sense. We consider the ring of integers $\mathcal{O}_K$ of K, i.e., the integral closure of $k[q]$ in K, and a q-difference algebra of the form

$$(4.1) \qquad \mathcal{A} = \mathcal{O}_K\left[x, \frac{1}{P(x)}, \frac{1}{P(qx)}, \frac{1}{P(q^2x)}, \cdots\right],$$

for some $P(x) \in \mathcal{O}_K[x]$, such that $q \in \mathcal{A}$. Then $\mathcal{A}$ is stable under the action of σ_q and we can consider a q-difference module $\mathcal{M} = (M, \Sigma_q)$ over $\mathcal{A}$. Remember that $\mathcal{M}_{K(x)} = (M_{K(x)} = M \otimes_{\mathcal{A}} K(x), \Sigma_q \otimes \sigma_q)$ is a q-difference module over $K(x)$ and that any q-difference module over $K(x)$ comes from a q-difference module over $\mathcal{A}$, for a convenient choice of $\mathcal{A}$.

We denote by ϕ_v the uniformizer of the cyclotomic place of $k(q)$ induced by $v \in \mathcal{C}_K$. The ring $\mathcal{A} \otimes_{\mathcal{O}_K} \mathcal{O}_K/(\phi_v)$ is not reduced in general, nevertheless it has a q-difference algebra structure and the results in [**DV02**, §2] apply again. Therefore we set:

DEFINITION 4.1. A q-difference module $\mathcal{M}$ over $\mathcal{A}$ has zero v-curvature (modulo ϕ_v) if the operator $\Sigma_q^{\kappa_v}$ induces the identity (or equivalently if the operator $\Delta_q^{\kappa_v}$, with $\Delta_q = \frac{\Sigma_q - 1}{(q-1)x}$, induces the zero operator) on the module $M \otimes_{\mathcal{A}} \mathcal{A}/(\phi_v)$.

Our main result is the following.

THEOREM 4.2. *A q-difference module $\mathcal{M}$ over $\mathcal{A}$ has zero v-curvature modulo ϕ_v, for almost all $v \in \mathcal{C}$, if and only if $\mathcal{M}$ becomes trivial over $K(x)$.*

REMARK 4.3. As proved in [**DV02**, Proposition 2.1.2], if $\Sigma_q^{\kappa_v}$ is the identity modulo ϕ_v then the q-difference module structure induced on $\mathcal{M} \otimes_{\mathcal{A}} \mathcal{A}/(\phi_v)$ is trivial.

As far as the proof of Theorem 4.2 is regarded, one implication is trivial. We will come back to the proof of the other implication in §4.3.

4.2. Regularity and triviality of the exponents

In this section, we are going to prove that a q-difference module is regular singular and has integral exponents if it has nilpotent reduction for sufficiently many cyclotomic places. We denote by π_v an uniformizer of $v \in \mathcal{C}$.

DEFINITION 4.4. We say that a q-difference module $\mathcal{M} = (M, \Sigma_q)$ over a q-difference $\mathcal{O}_K$-algebra $\mathcal{A}$, as above, has nilpotent v-curvature modulo π_v, or simply that it has nilpotent reduction modulo π_v, if the linear operator $\Sigma_q^{\kappa_v} : M \otimes_{\mathcal{A}} \mathcal{A}/(\pi_v) \longrightarrow M \otimes_{\mathcal{A}} \mathcal{A}/(\pi_v)$ is unipotent (or equivalently if $\Delta_q^{\kappa_v}$ is nilpotent. See [**DV02**, §2]).

We prove the following result:

PROPOSITION 4.5.

(1) If a *q-difference* module $\mathcal{M}$ over $\mathcal{A}$ has nilpotent v-curvature modulo π_v, for infinitely many $v \in \mathcal{C}$, then it is regular singular.
(2) Let $\mathcal{M}$ be a *q-difference* module over $\mathcal{A}$. If there exists an infinite set of positive primes $\wp \subset \mathbb{Z}$ such that $\mathcal{M}$ has nilpotent v-curvature modulo π_v, for all $v \in \mathcal{C}$, such that $\kappa_v \in \wp$, then $\mathcal{M}_{K((x))}$ is trivial.

PROOF. The proof of Proposition 3.10 applies word by word to this case, until the argument showing that A_0 is diagonalisable. To conclude with Proposition 2.4, one has to show that the eigenvalues of A_0 are in $q^{\mathbb{Z}}$. Let $\widetilde{K}$ be a finite extension of K in which we can find all the eigenvalues of A_0. Then any eigenvalue $\alpha \in \widetilde{K}$ of A_0 has the property that $\alpha^{\kappa_v} = 1$ modulo w, for all $w \in \mathcal{C}_{\widetilde{K}}$, $w|v$ and v satisfies the assumptions. In other words, the reduction modulo w of an eigenvalue α of A_0 belongs to the multiplicative cyclic group generated by the reduction of q modulo π_v.

To end the proof, we are reduced to prove the proposition below. □

PROPOSITION 4.6. *Let k be a perfect field, $K/k(q)$ be a finite extension and $\wp \subset \mathbb{Z}$ be an infinite set of positive primes. For any $v \in \mathcal{C}$, let κ_v be the order of q modulo π_v, as a root of unity.*

If $\alpha \in K$ is such that $\alpha^{\kappa_v} \equiv 1$ modulo π_v, for all $v \in \mathcal{C}$ such that $\kappa_v \in \wp$, then $\alpha \in q^{\mathbb{Z}}$.

REMARK 4.7. Let $K = \mathbb{Q}(\widetilde{q})$, with $\widetilde{q}^r = q$, for some integer $r > 1$. If $\widetilde{q}$ is an eigenvalue of A_0 we would be asking that for infinitely many positive primes $\ell \in \mathbb{Z}$ there exists a primitive root of unity $\zeta_{r\ell}$ of order $r\ell$, which is also a root of unity of order ℓ. Of course, this cannot be true, unless $r = 1$.

4.2.1. Proof of Proposition 4.6. We denote by k_0 either the field of rational numbers $\mathbb{Q}$, if the characteristic of k is zero, or the field with p elements $\mathbb{F}_p$, if the characteristic of k is $p > 0$. First of all, let us suppose that k is a finite perfect extension of k_0 of degree d and fix an embedding $k \hookrightarrow \overline{k}$ of k in its algebraic closure $\overline{k}$. In the case of a rational function $\alpha = f(q) \in k(q)$, Proposition 4.6 is a consequence of the following lemma:

LEMMA 4.8. *Let k be a perfect field, $[k : k_0] = d < \infty$ and let $f(q) \in k(q)$ be non-zero rational function. If there exists an infinite set of positive primes $\wp \subset \mathbb{Z}$ with the following property:*

> *for any $\ell \in \wp$ there exists a primitive root of unity ζ_ℓ of order ℓ such that $f(\zeta_\ell)$ is a root of unity of order ℓ,*

then $f(q) \in q^{\mathbb{Z}}$.

REMARK 4.9. If $k = \mathbb{C}$ and $y - f(q)$ is irreducible in $\mathbb{C}[q, y]$, the result can be deduced from [**Lan83**, Chapter 8, Theorem 6.1], whose proof uses Bézout theorem. We give here a totally elementary proof, that holds also in positive characteristic.

Proposition 4.6 can be rewritten in the language of rational dynamic. We denote by μ_ℓ the group of root of unity of order ℓ. The following assertions are equivalent:
(1) $f(q) \in k(q)$ satisfies the assumptions of Lemma 4.8.
(2) There exist infinitely many $\ell \in \mathbb{N}$ such that the group μ_ℓ of roots of unity of order ℓ verifies $f(\mu_\ell) \subset \mu_\ell$.
(3) $f(q) \in q^{\mathbb{Z}}$.

(4) The Julia set of f is the unit circle.

As it was pointed out to us by C. Favre, the equivalence between the last two assumptions is a particular case of [**Zdu97**], while the equivalence between the second and the fourth assumption can be deduced from [**FRL06**] or [**Aut01**].

PROOF. Let $f(q) = \frac{P(q)}{Q(q)}$, with $P = \sum_{i=0}^{D} a_i q^i, Q = \sum_{i=0}^{D} b_i q^i \in k[q]$ coprime polynomials of degree less than or equal to D, and let ℓ be a prime in $\wp$, such that:

- $f(\zeta_\ell) \in \mu_\ell$;
- $2D < \ell - 1$.

Moreover, since $\wp$ is infinite, we can chose $\ell >> 0$ so that the extensions k and $k_0(\mu_\ell)$ are linearly disjoint over k_0. Since k is perfect, this implies that the minimal polynomial of the primitive ℓ-th root of unity ζ_ℓ over k is $\chi(X) = 1 + X + \cdots + X^{\ell-1}$. Now let $\kappa \in \{0, \ldots, \ell-1\}$ be such that $f(\zeta_\ell) = \zeta_\ell^\kappa$, i.e.,

$$\sum_{i=0}^{D} a_i \zeta_\ell^i = \sum_{i=0}^{D} b_i \zeta_\ell^{i+\kappa}.$$

We consider the polynomial $H(q) = \sum_{i=0}^{D} a_i q^i - \sum_{j=\kappa}^{D+\kappa} b_{j-\kappa} q^j$ and distinguish three cases:

(1) If $D + \kappa < \ell - 1$, then $H(q)$ has ζ_ℓ as a zero and has degree strictly inferior to $\ell - 1$. Necessarily $H(q) = 0$. Since $\kappa > D$ would imply $f = 0$, we have $\kappa \le D$. Thus we have

$$a_0 = a_1 = \cdots = a_{\kappa-1} = b_{D+1-\kappa} = \cdots = b_D = 0 \quad \text{and} \quad a_i = b_{i-\kappa} \text{ for } i = \kappa, \ldots, D,$$

which implies $f(q) = q^\kappa$.

(2) If $D + \kappa = \ell - 1$ then $H(q)$ is a k-multiple of $\chi(q)$ and therefore all the coefficients of $H(q)$ are all equal. Notice that the inequality $D + \kappa \ge \ell - 1$ forces κ to be strictly bigger than D, in fact otherwise one would have $\kappa + D \le 2D < \ell - 1$. For this reason the coefficients of $H(q)$ of the monomials $q^{D+1}, \ldots, q^\kappa$ are all equal to zero. Thus

$$a_0 = a_1 = \cdots = a_D = b_0 = \cdots = b_D = 0$$

and therefore $f = 0$ against the assumptions. So the case $D + \kappa = l - 1$ cannot occur.

(3) If $D + \kappa > \ell - 1$, then $\kappa > D > D + \kappa - \ell$, since $\kappa > D$ and $\kappa - \ell < 0$. In this case we shall rather consider the polynomial $\widetilde{H}(q)$ defined by:

$$\widetilde{H}(q) = \sum_{i=0}^{D} a_i q^i - \sum_{i=\kappa}^{\ell-1} b_{i-\kappa} q^i - \sum_{i=0}^{D+\kappa-\ell} b_{i+\ell-\kappa} q^i.$$

Notice that $H(\zeta_\ell) = \widetilde{H}(\zeta_\ell) = 0$ and that $\widetilde{H}(q)$ has degree smaller or equal to $\ell - 1$. As in the previous case, $\widetilde{H}(q)$ is a k-multiple of $\chi(q)$. We get

$$b_j = 0 \text{ for } j = 0, \ldots, \ell - 1 - \kappa$$

and

$$a_0 - b_{\ell-\kappa} = \cdots = a_{D+\kappa-\ell} - b_D = a_{D+\kappa-\ell+1} = \cdots = a_D = 0.$$

We conclude that $f(q) = q^{\kappa-\ell}$.

This ends the proof. □

4.2. REGULARITY AND TRIVIALITY OF THE EXPONENTS

We are going to deduce Proposition 4.6 from Lemma 4.8 in two steps: first of all we are going to show that we can drop the assumption that $[k : k_0]$ is finite and then that one can always reduce to the case of a rational function.

LEMMA 4.10. *Lemma 4.8 holds if k/k_0 is a finitely generated (not necessarily algebraic) extension.*

REMARK 4.11. Since $f(q) \in k(q)$, replacing k by the field generated by the coefficients of f over k_0, we can always assume that k/k_0 is finitely generated.

PROOF. Let $\widetilde{k}$ be the algebraic closure of k_0 in k and let k' be an intermediate field of $k/\widetilde{k}$, such that $f(q) \in k'(q) \subset k(q)$ and that $k'/\widetilde{k}$ has minimal transcendence degree ι. We suppose that $\iota > 0$, to avoid the situation of Lemma 4.8. So let $a_1, \ldots, a_\iota$ be transcendence basis of $k'/\widetilde{k}$ and let $k'' = \widetilde{k}(a_1, \ldots, a_\iota)$. If $k'/\widetilde{k}$ is purely transcendental, i.e., if $k' = k''$, then $f(q) = P(q)/Q(q)$, where $P(q)$ and $Q(q)$ can be written in the form:

$$P(q) = \sum_i \sum_{\underline{j}} \alpha_{\underline{j}}^{(i)} a_{\underline{j}} q^i \quad \text{and} \quad Q(q) = \sum_i \sum_{\underline{j}} \beta_{\underline{j}}^{(i)} a_{\underline{j}} q^i,$$

with $\underline{j} = (j_1, \ldots, j_\iota) \in \mathbb{Z}_{\geq 0}^\iota$, $a_{\underline{j}} = a_{j_i} \cdots a_{j_\iota}$ and $\alpha_{\underline{j}}^{(i)}, \beta_{\underline{j}}^{(i)} \in \widetilde{k}$. If we reorganize the terms of P and Q so that

$$P(q) = \sum_{\underline{j}} a_{\underline{j}} D_{\underline{j}}(q) \quad \text{and} \quad Q(q) = \sum_{\underline{j}} a_{\underline{j}} C_{\underline{j}}(q),$$

we conclude that the assumption $f(\zeta_\ell) \subset \mu_\ell$ for infinitely many primes ℓ implies that $f_{\underline{j}} = \frac{D_{\underline{j}}}{C_{\underline{j}}}$ is a rational function with coefficients in $\widetilde{k}$ satisfying the assumptions of Lemma 4.8. Moreover, since the $f_{\underline{j}}$'s take the same values at infinitely many roots of unity, they are all equal. Finally, we conclude that $f_{\underline{j}}(q) = q^d$ for any $\underline{j}$ and hence that $f = q^d \frac{\sum \alpha_{\underline{j}}}{\sum \alpha_{\underline{j}}} = q^d$.

Now let us suppose that $k' = k''(b)$ for some primitive element b, algebraic over k'', of degree e. Then once again we write $f(q) = P(q)/Q(q)$, with:

$$P(q) = \sum_i \sum_{h=0}^{e-1} \alpha_{i,h} b^h q^i \quad \text{and} \quad Q(q) = \sum_i \sum_{h=0}^{e-1} \beta_{i,h} b^h q^i,$$

with $\alpha_{i,h}, \beta_{i,h} \in k''$. Again we conclude that $\frac{\sum_i \alpha_{i,h} q^i}{\sum_i \beta_{i,h} q^i} = q^d$ for any $h = 0, \ldots, e-1$, and hence that $f(q) = q^d$. □

END OF THE PROOF OF PROPOSITION 4.6. Let $\widetilde{K} = k(q, f) \subset K$. If the characteristic of k is p, replacing f by a p^n-th power of f, we can suppose that $\widetilde{K}/k(q)$ is a Galois extension. So we set:

$$y = \prod_{\varphi \in Gal(\widetilde{K}/k(q))} f^\varphi \in k(q).$$

For infinitely many $v \in \mathcal{C}_{k(q)}$ such that κ_v is a prime, we have $f^{\kappa_v} \equiv 1$ modulo w, for any $w|v$. Since $Gal(\widetilde{K}/K)$ acts transitively over the set of places $w \in \mathcal{C}_{\widetilde{K}}$ such that $w|v$, this implies that $y^{\kappa_v} \equiv 1$ modulo π_v. Then Lemmas 4.10 and 4.8 allow us to conclude that $y \in q^{\mathbb{Z}}$. This proves that we are in the following situation: f is an algebraic function such that $|f|_w = 1$ for any $w \in \mathcal{P}_{\widetilde{K},f}$ and that $|f|_w \neq 1$

for any $w \in \mathcal{P}_{\widetilde{K},\infty}$. We conclude that $f = cq^{s/r}$ for some non-zero integers s, r and some constant c in a finite extension of k. Since $f^{\kappa_v} \equiv 1$ modulo w, for all $w \in \mathcal{C}_{\widetilde{K}}$ such that $\kappa_v \in \wp$, we finally obtain that $r = 1$ and $c = 1$. $\square$

4.3. Proof of Theorem 4.2

Under the assumption of Theorem 4.2, Proposition 4.5 implies that the q-difference module $\mathcal{M}$ becomes trivial over $K((x))$. To conclude we need to show the following proposition:

PROPOSITION 4.12. *If a q-difference module $\mathcal{M}$ over $\mathcal{A}$ has zero v-curvature modulo ϕ_v, for almost all $v \in \mathcal{C}$, then there exists a basis $\underline{e}$ of $M_{K(x)}$ over $K(x)$ such that the associated q-difference system has a formal fundamental solution matrix $Y(x) \in \mathrm{GL}_\nu(K((x)))$, which is the Taylor expansion at 0 of a matrix in $\mathrm{GL}_\nu(K(x))$, i.e., $\mathcal{M}$ becomes trivial over $K(x)$.*

REMARK 4.13. This is the only part of the proof of Theorem 4.2 where we need to suppose that the v-curvatures are zero modulo ϕ_v, for almost all $v \in \mathcal{C}$.

PROOF. (*cf.* [**DV02**, Proposition 8.2.1]) Let $\underline{e}$ be a basis of M over $K(x)$. Applying a basis change with coefficients in $K\left[x, \frac{1}{x}\right]$, we can actually suppose that $\Sigma_q \underline{e} = \underline{e} A(x)$, where $A(x) \in \mathrm{GL}_\nu(K(x))$ has no pole at 0 and $A(0)$ is the identity matrix. In the notation of §1.3, the recursive relation defining the matrices $G_n(x)$ implies that they have no pole at 0. This means that $Y(x) := \sum_{n \geq 0} G_{[n]}(0) x^n$ is a fundamental solution matrix of the q-difference system associated to $\mathcal{M}_{K(x)}$ with respect to the basis $\underline{e}$.

We recall the definition of the Gauss norm associated to an ultrametric norm $v \in \mathcal{P}$:

$$\text{for any } \frac{\sum a_i x^i}{\sum b_j x^j} \in K(x), \quad \left|\frac{\sum a_i x^i}{\sum b_j x^j}\right|_{v,Gauss} = \frac{\sup |a_i|_v}{\sup |b_j|_v}.$$

We have:

LEMMA 4.14. *Let $v \in \mathcal{C}_K$. We assume that $|G_1(x)|_{v,Gauss} \leq 1$. Then the following assertions are equivalent:*
  (1) *The module $\mathcal{M} = (M, \Sigma_q)$ has zero v-curvature modulo ϕ_v.*
  (2) *For any positive integer n, we have $|G_{[n]}|_{v,Gauss} \leq 1$.*

REMARK 4.15. Let k_v be the residue field of K modulo v and q_v the reduction of q in k_v, which is defined for almost all $v \in \mathcal{C}$. According to [**Har10**, §3], the second assertion of the lemma above can be rewritten as: $\mathcal{M}_{k_v(x)}$ has a natural structure of iterated q_v-difference module.

PROOF OF LEMMA 4.14. The only non-trivial implication is "1 $\Rightarrow$ 2" whose proof is quite similar to [**DV02**, Lemma 5.1.2]. The Leibniz Formula for d_q and Δ_q implies that:

$$G_{(n+1)\kappa_v} = \sum_{i=0}^{\kappa_v} \binom{\kappa_v}{i}_q \sigma_q^{\kappa_v - i}(d_q^i(G_{n\kappa_v})) G_{\kappa_v - i}.$$

If $\mathcal{M}$ has zero v-curvature modulo ϕ_v then $|G_{\kappa_v}|_{v,Gauss} \leq |\phi_v|_v$. One obtains recursively that $|G_m|_{v,Gauss} \leq |\phi_v|_v^{\left[\frac{m}{\kappa_v}\right]}$, where we have denoted by $[a]$ the integral part of $a \in \mathbb{R}$, i.e., $[a] = max\{n \in \mathbb{Z} : n \leq a\}$. Since $|[\kappa_v]_q|_v = |\phi_v|_v$ and $|[m]_q^!|_v = |\phi_v|_v^{\left[\frac{m}{\kappa_v}\right]}$, we conclude that:

$$\left|\frac{G_m}{[m]_q^!}\right|_{v,Gauss} \leq 1. \tag{4.2}$$

This ends the proof of the lemma. $\square$

We go back to the proof of Proposition 4.12. The entries of $Y(x) = \sum_{n \geq 0} G_{[n]}(0)x^n$ verify the following properties:

- For any $v \in \mathcal{P}_\infty$, the matrix $Y(x)$ is analytic at 0 and has infinite v-adic radius of meromorphy (see Proposition 1.9).
- Since $|[n]_q|_{v,Gauss} = 1$ for any non-cyclotomic place $v \in \mathcal{P}_f$, we have $\left|G_{[m]}(x)\right|_{v,Gauss} \leq 1$, for almost all $v \in \mathcal{P}_f \setminus \mathcal{C}$. For the finitely many $v \in \mathcal{P}_f$ such that $|G_1(x)|_{v,Gauss} > 1$, there exists a constant $C > 0$ such that $\left|G_{[m]}(x)\right|_{v,Gauss} \leq C^m$, for any positive integer m.
- For almost all $v \in \mathcal{C}$ and all positive integer m, $\left|G_{[m]}(x)\right|_{v,Gauss} \leq 1$ (cf. Lemma 4.14), while for the remaining finitely many $v \in \mathcal{C}$ there exists a constant $C > 0$ such that $\left|G_{[m]}(x)\right|_{v,Gauss} \leq C^m$ for any positive integer m.

This implies that:

$$\limsup_{m \to \infty} \frac{1}{m} \sum_{v \in \mathcal{P}} \log^+ \left|G_{[m]}(x)\right|_{v,Gauss} < \infty.$$

Since for almost all v we have $\left|G_{[m]}(0)\right|_v \leq \left|G_{[m]}(x)\right|_{v,Gauss}$, we conclude that $Y(x)$ is a matrix with rational entries applying a simplified form of the Borel-Dwork criteria for function fields (see [DV02, Propositions 8.2.1 and 8.4.1]),[3] which is itself a simplification of the more general criteria [**And04**, Theorem 5.4.3]. We are omitting the details. $\square$

4.4. Link with iterative q-difference equations

We denote by k_v the residue field of K with respect to a place $v \in \mathcal{P}$ and by q_v the image of q in k_v, which is defined for all places $v \in \mathcal{P}$. For almost all $v \in \mathcal{P}_f$ we can consider the $k_v(x)$-vector space $M_{k_v(x)} = M \otimes_\mathcal{A} k_v(x)$, with the structure induced by Σ_q. In this way, for almost all $v \in \mathcal{P}$, we obtain a q_v-difference module $\mathcal{M}_{k_v(x)} = (M_{k_v(x)}, \Sigma_{q_v})$ over $k_v(x)$,

In the framework of iterative q-difference equations [**Har10**], Theorem 4.2 is equivalent to the following statement, which is a q-analogue of the conjecture stated at the very end of [**MvdP03**]:

COROLLARY 4.16. *For a q-difference module $\mathcal{M}$ over $\mathcal{A}$ the following statement are equivalent:*

(1) *The q-difference module $\mathcal{M}$ over $\mathcal{A}$ becomes trivial over $K(x)$;*
(2) *It induces an iterative q_v-difference structure over $\mathcal{M}_{k_v(x)}$, for almost all $v \in \mathcal{C}$;*
(3) *It induces a trivial iterative q_v-difference structure over $\mathcal{M}_{k_v(x)}$, for almost all $v \in \mathcal{C}$.*

---

[3]The simplification comes from the fact, in this setting, that there are no archimedean norms.

PROOF. The equivalence $1 \Leftrightarrow 2$ is a consequence of Lemma 4.14 and Theorem 4.2, while the implication $3 \Rightarrow 2$ is tautological.

Let us prove that $1 \Rightarrow 3$. If the q-difference module $\mathcal{M}$ becomes trivial over $K(x)$, then there exist an $\mathcal{A}$-algebra $\mathcal{A}'$, of the form (4.1), obtained from $\mathcal{A}$ inverting a polynomial and its q-iterates, and a basis $\underline{e}$ of $M \otimes_\mathcal{A} \mathcal{A}'$ over $\mathcal{A}'$, such that the associated q-difference system is $\sigma_q(Y) = Y$. Therefore, for almost all $v \in \mathcal{C}$, $\mathcal{M}$ induces an iterative q_v-difference module $\mathcal{M}_{k_v(x)}$ whose iterative q_v-difference equations are given by $\frac{d_{q_v}^{\kappa_v}}{[\kappa_v]!_{q_v}}(Y) = 0$ (cf. [**Har10**, Proposition 3.17]). □

CHAPTER 5

A unified statement

Let K be a field, $q \in K$, $q \neq 0, 1$ be a fixed element. If follows from Proposition 1.4 that we can suppose that K is finitely generated over the prime field. Let $\mathcal{M}_{K(x)} = (M_{K(x)}, \Sigma_q)$ be a q-difference module over $K(x)$. We recall the following notations:

($\mathcal{A}lg$) *If q is algebraic over $\mathbb{Q}$, but not a root of unity*, we are in the following situation. We call Q the algebraic closure of $\mathbb{Q}$ inside K, $\mathcal{O}_Q$ the ring of integer of Q, $\mathcal{C}$ the set of finite places v of Q and $\pi_v \in \mathcal{O}_Q$ a v-adic uniformizer. For almost all finite place v of Q, the following objects are well-defined: the order κ_v, as a root of unity, of the reduction of q modulo π_v and the positive integer power ϕ_v of π_v, such that $\phi_v^{-1}(1 - q^{\kappa_v})$ is a unit of $\mathcal{O}_Q$. The field K has the form $Q(\underline{a}, b)$, where $\underline{a} = (a_1, \ldots, a_r)$ is a transcendence basis of K/Q and b is a primitive element of the algebraic extension $K/Q(\underline{a})$. Choosing conveniently the set of generators $\underline{a}, b$ and $P(x) \in \mathcal{O}_Q[\underline{a}, b, x]$, we can always find a q-difference algebra $\mathcal{A}$ of the form

(5.1) $$\mathcal{A} = \mathcal{O}_Q\left[\underline{a}, b, x, \frac{1}{P(x)}, \frac{1}{P(qx)}, \ldots\right]$$

and a Σ_q-stable $\mathcal{A}$-lattice M of $\mathcal{M}_{K(x)}$, so that we can consider the $\mathcal{A}/(\phi_v)$-linear operator

$$\Sigma_q^{\kappa_v} : M \otimes_\mathcal{A} \mathcal{A}/(\phi_v) \longrightarrow M \otimes_\mathcal{A} \mathcal{A}/(\phi_v),$$

that we have called the v-curvature of $\mathcal{M}_{K(x)}$-modulo ϕ_v. Notice that $\mathcal{O}_Q/(\phi_v)$ is not an integral domain in general.

($\mathcal{T}rans$) *If q is transcendental over the prime field of K*, then there exists a subfield k of K such that K is a finite extension of $k(q)$. We denote by $\mathcal{C}$ the set of places of K that extend the places of $k(q)$, associated to irreducible polynomials ϕ_v of $k[q]$, that vanish at roots of unity. Let κ_v be the order of the roots of ϕ_v. Let $\mathcal{O}_K$ be the integral closure of $k[q]$ in K. Choosing conveniently $P(x) \in \mathcal{O}_K[x]$, we can always find a q-difference algebra $\mathcal{A}$ of the form:

(5.2) $$\mathcal{A} = \mathcal{O}_K\left[x, \frac{1}{P(x)}, \frac{1}{P(qx)}, \ldots\right]$$

and a Σ_q-stable $\mathcal{A}$-lattice M of $\mathcal{M}_{K(x)}$, so that we can consider the $\mathcal{A}/(\phi_v)$-linear operator

$$\Sigma_q^{\kappa_v} : M \otimes_\mathcal{A} \mathcal{A}/(\phi_v) \longrightarrow M \otimes_\mathcal{A} \mathcal{A}/(\phi_v),$$

that we have also called the v-curvature of $\mathcal{M}_{K(x)}$-modulo ϕ_v. Notice that, once again, $\mathcal{O}_K/(\phi_v)$ is not an integral domain in general.

($\mathcal{R}oot$) *If q is a primitive root of unity of order κ*, we define $\mathcal{C}$ to be the set containing only the trivial valuation v on K, $\phi_v = 0$ and $\kappa_v = \kappa$. Then there exists a polynomial $P(x) \in K[x]$ such that the algebra $\mathcal{A} = K\left[x, \frac{1}{P(x)}, \frac{1}{P(qx)}, \dots\right]$ is σ_q-stable and there exists a Σ_q-stable $\mathcal{A}$-lattice M of $\mathcal{M}_{K(x)}$, so that we can consider the $\mathcal{A}/(\phi_v)$-linear operator

$$\Sigma_q^{\kappa_v} : M \otimes_{\mathcal{A}} \mathcal{A}/(\phi_v) \longrightarrow M \otimes_{\mathcal{A}} \mathcal{A}/(\phi_v),$$

that we will call the v-curvature of $\mathcal{M}_{K(x)}$-modulo ϕ_v. Notice that this is simply the κ-th iterate of Σ_q, namely $\Sigma_q^\kappa : M \longrightarrow M$.

Then the main result of the first part of this work is:

THEOREM 5.1. *A q-difference module $\mathcal{M}_{K(x)} = (M_{K(x)}, \Sigma_q)$ over $K(x)$ is trivial if and only if there exist an algebra $\mathcal{A}$, as above, and a Σ_q-stable $\mathcal{A}$-lattice M of $M_{K(x)}$ such that the map*

$$\Sigma_q^{\kappa_v} : M \otimes_{\mathcal{A}} \mathcal{A}/(\phi_v) \longrightarrow M \otimes_{\mathcal{A}} \mathcal{A}/(\phi_v),$$

is the identity, for any v in a cofinite non-empty subset of $\mathcal{C}$.

In the case $(\mathcal{A}lg)$ we can take $\mathcal{C}$ to be a set of finite places of Q of density 1, depending on $\mathcal{M}_{K(x)}$.

PROOF. So the statement above coincides with Proposition 3.1 if q is a root of unity, and with Theorem 3.8 if q is algebraic, but not a root of unity. Finally, to deduce the third case from Theorem 4.2, it is enough to remark that we can replace k by its perfect closure. $\square$

Of course, for a given module $\mathcal{M}_{K(x)}$ we can always find a q-difference algebra $\mathcal{A}$ as above and a q-difference module M over $\mathcal{A}$ such that $M \otimes_{\mathcal{A}} K(x) \cong \mathcal{M}_{K(x)}$. Also, if the statement above is true for a choice of $\mathcal{A}$ and one q-difference module M over $\mathcal{A}$, then it is true for all choice of $\mathcal{A}$ and of M. In the following chapters, we will use this fact implicitly.

Part 3

Intrinsic Galois groups

CHAPTER 6

The intrinsic Galois group

6.1. Definition and first properties

Let $\mathcal{F}$ be a q-difference field and $\mathcal{M}_\mathcal{F} = (M_\mathcal{F}, \Sigma_q)$ be a q-difference module of rank ν over $\mathcal{F}$, in the sense of Chapter 1. We can consider the family $Constr_\mathcal{F}(\mathcal{M}_\mathcal{F})$ of q-difference modules containing $\mathcal{M}_\mathcal{F}$ and closed under direct sum, tensor product, dual, symmetric and antisymmetric products (see §1.1.1). We will denote by $Constr_\mathcal{F}(\mathcal{M}_\mathcal{F})$ the collection of constructions of linear algebra of the $\mathcal{F}$-vector space $M_\mathcal{F}$, i.e., the collection of underlying $\mathcal{F}$-vector spaces of the family $Constr_\mathcal{F}(\mathcal{M}_\mathcal{F})$. We denote by $\mathrm{GL}(M_\mathcal{F})$, the group scheme that attach to any $\mathcal{F}$-algebra S the group of S-linear automorphisms of $M_\mathcal{F} \otimes S$. This group scheme acts naturally, by functoriality, on any element of $Constr_\mathcal{F}(\mathcal{M}_\mathcal{F})$, after a scalar extension from $\mathcal{F}$ to S.

DEFINITION 6.1. *The intrinsic Galois group*[1] $Gal(\mathcal{M}_\mathcal{F})$ *of* $\mathcal{M}_\mathcal{F}$ *is the subgroup of* $\mathrm{GL}(M_\mathcal{F})$ *which stabilizes all the q-difference submodules over $\mathcal{F}$ of any object in* $Constr_\mathcal{F}(\mathcal{M}_\mathcal{F})$.

In the definitions above and below, the term "stabilizer" has to be understood in the functorial sense of [**DG70**, II.1.36]. For instance, $Gal(\mathcal{M}_\mathcal{F})$ is a functor from the category of $\mathcal{F}$-algebras to the category of groups, that associates to any $\mathcal{F}$-algebra S the subgroup of $\mathrm{GL}(M_\mathcal{F})(S)$, the S-points of the $\mathcal{F}$-group scheme $\mathrm{GL}(M_\mathcal{F})$, that stabilizes $\mathcal{N}_\mathcal{F} \otimes S$, for all the q-difference submodules $\mathcal{N}_\mathcal{F}$ over $\mathcal{F}$ of any object in $Constr_\mathcal{F}(\mathcal{M}_\mathcal{F})$. By [**DG70**, II.1.36], this functor is representable and thus defines a group scheme over $\mathcal{F}$.

Notice that in positive characteristic p, the group $Gal(\mathcal{M}_\mathcal{F})$ is not necessarily reduced. An easy example is given by the equation $y(qx) = q^{1/p}y(x)$, whose intrinsic Galois group is μ_p (*cf.* [**vdPR07**, §7]).

REMARK 6.2. The group $Gal(\mathcal{M}_\mathcal{F})$ can be interpreted in a Tannakian framework as the group of tensor automorphisms of the forgetful functor of the full Tannakian subcategory generated by $\mathcal{M}_\mathcal{F}$ in $Diff(\mathcal{F}, \sigma_q)$ (see [**And01**, §3.2.2.2.]). This Tannakian point of view allows to compare the different notions of Galois groups attached to q-difference modules.

REMARK 6.3. We recall that the Chevalley theorem, that also holds for non-reduced groups (*cf.* [**DG70**, II, §2, n.3, Corollary 3.5]), ensures that $Gal(\mathcal{M}_\mathcal{F})$ can be defined as the stabilizer of a rank one submodule (which is not necessarily a q-difference module) of a q-difference module contained in an algebraic construction of $\mathcal{M}_\mathcal{F}$. Nevertheless, it is possible to find a rank one q-difference module that defines $Gal(\mathcal{M}_\mathcal{F})$ as its stabilizer. In fact the noetherianity of $\mathrm{GL}(M_\mathcal{F})$ implies that

---

[1]In the literature, the intrinsic Galois group is also called the generic Galois group of $\mathcal{M}_\mathcal{F}$.

$Gal(\mathcal{M}_\mathcal{F})$ is defined as the stabilizer of a finite family of q-difference submodules $\mathcal{W}_\mathcal{F}^{(i)} = (W_\mathcal{F}^{(i)}, \Sigma_q)$ contained in some constructions of linear algebra $\mathcal{M}_\mathcal{F}^{(i)}$ of $\mathcal{M}_\mathcal{F}$. It follows that the line

$$L_\mathcal{F} = \wedge^{\dim \oplus_i W_\mathcal{F}^{(i)}} \left(\bigoplus_i W_\mathcal{F}^{(i)} \right) \subset \wedge^{\dim \oplus_i W_\mathcal{F}^{(i)}} \left(\bigoplus_i \mathcal{M}_\mathcal{F}^{(i)} \right)$$

is a q-difference module and defines $Gal(\mathcal{M}_\mathcal{F})$ as a stabilizer (*cf.* [**Kat82**, proof of Proposition 9]).

NOTATION 6.4. In the sequel, we will use the notation $Stab(W_\mathcal{F}^{(i)}, i)$ to say that a group is the stabilizer of the family of vector spaces $\{W_\mathcal{F}^{(i)}\}_i$.

6.2. Arithmetic characterization of the intrinsic Galois group

From now on, we consider the particular case $\mathcal{F} = K(x)$, with the notation introduced in Chapter 5. Let G be a $K(x)$-subgroup scheme of $GL(M_{K(x)})$, such that $G = Stab(L_{K(x)})$ for some line $L_{K(x)}$ contained in a construction of linear algebra $\mathcal{W}_{K(x)}$ of $\mathcal{M}_{K(x)}$. Let M be a Σ_q-stable $\mathcal{A}$-lattice of $M_{K(x)}$ for some q-difference algebra $\mathcal{A}$. Up to enlarging $\mathcal{A}$, one finds an $\mathcal{A}$-lattice L of $L_{K(x)}$ and an $\mathcal{A}$-lattice W of $W_{K(x)}$. The latter is the underlying space of a q-difference module $\mathcal{W} = (W, \Sigma_q)$ over $\mathcal{A}$.

DEFINITION 6.5. Let $\widetilde{\mathcal{C}}$ be a cofinite non-empty subset of $\mathcal{C}$ and $(\Lambda_v)_{v \in \widetilde{\mathcal{C}}}$ be a family of invertible $\mathcal{A}/(\phi_v)$-linear operators acting on $M \otimes_\mathcal{A} \mathcal{A}/(\phi_v)$, for any $v \in \widetilde{\mathcal{C}}$, respectively. We say that the $K(x)$-subgroup scheme $G \subset GL(M_{K(x)})$ contains the operators Λ_v modulo ϕ_v for almost all $v \in \mathcal{C}$ if for almost all, and at least one, $v \in \widetilde{\mathcal{C}}$ the operator Λ_v stabilizes $L \otimes_\mathcal{A} \mathcal{A}/(\phi_v)$ inside $W \otimes_\mathcal{A} \mathcal{A}/(\phi_v)$:

$$\Lambda_v \in Stab_{\mathcal{A}/(\phi_v)}(L \otimes_\mathcal{A} \mathcal{A}/(\phi_v)).$$

REMARK 6.6. As in [**DV02**, 10.1.2], one can prove that the definition above is independent of the choice of $\mathcal{A}$, M and $L_{K(x)}$.

NOTATION 6.7. From now on, we will always use the phrase "for almost all" to mean "for almost all, and at least one". In this way the statements will be correct even in the case ($\mathcal{R}oot$) (see Chapter 5).

The main result of this section is the following:

THEOREM 6.8. *The $K(x)$-group scheme $Gal(\mathcal{M}_{K(x)})$ is the smallest $K(x)$-subgroup scheme of $GL(M_{K(x)})$, that contains the operators $\Sigma_q^{\kappa_v}$ modulo ϕ_v for almost all $v \in \mathcal{C}$.*

REMARK 6.9.
- The noetherianity of $GL(M_{K(x)})$ implies that the smallest $K(x)$-subgroup scheme of $GL(M_{K(x)})$ that contains the operators $\Sigma_q^{\kappa_v}$ modulo ϕ_v, for almost all $v \in \mathcal{C}$, is well-defined. Theorem 6.8 has been proved in [**Hen96**, Chapter 6] when q is a root of unity and in [**DV02**] when q is algebraic and K is a number field.
- Under the assumption ($\mathcal{A}lg$) (see Chapter 5), the statement above is still true if we replace $\mathcal{C}$ by a set of finite places of Q of density 1. This remark applies to all statements in this and the next chapter.

A part of Theorem 6.8 is easy to prove:

LEMMA 6.10. *The $K(x)$-group scheme $Gal(\mathcal{M}_{K(x)})$ contains the operators $\Sigma_q^{\kappa_v}$ modulo ϕ_v for almost all $v \in \mathcal{C}$.*

PROOF. The statement follows immediately from the fact that $Gal(\mathcal{M}_{K(x)})$ can be defined as the stabilizer of a rank one q-difference module, that is contained in a construction of linear algebra of $\mathcal{M}_{K(x)}$, which is *a fortiori* stable under the action of $\Sigma_q^{\kappa_v}$. □

COROLLARY 6.11 (Lemma 9.34 in [**DV02**]). *$Gal(\mathcal{M}_{K(x)}) = \{1\}$ if and only if $\mathcal{M}_{K(x)}$ is a trivial q-difference module.*

Now we are ready to give the proof of Theorem 6.8, whose main ingredient is Theorem 5.1. The argument is inspired by [**Kat82**, §X].

PROOF OF THEOREM 6.8. Lemma 6.10 says that $Gal(\mathcal{M}_{K(x)})$ contains the smallest $K(x)$-subgroup scheme G of $GL(\mathcal{M}_{K(x)})$, that contains the operator $\Sigma_q^{\kappa_v}$ modulo ϕ_v for almost all $v \in \mathcal{C}$. Let $L_{K(x)}$ be a line contained in some construction of linear algebra of $\mathcal{M}_{K(x)}$, that defines G as a stabilizer. Then there exists a smaller q-difference module $\mathcal{W}_{K(x)}$ over $K(x)$ that contains $L_{K(x)}$. Let L and $\mathcal{W} = (W, \Sigma_q)$ be the associated $\mathcal{A}$-modules. Any generator m of L as an $\mathcal{A}$-module is a cyclic vector for $\mathcal{W}$ and the operator $\Sigma_q^{\kappa_v}$ acts on $W \otimes_\mathcal{A} \mathcal{A}/(\phi_v)$ with respect to the basis induced by the cyclic basis generated by m via a diagonal matrix. Because of the definition of the q-difference structure on the dual module $\mathcal{W}^*$ of $\mathcal{W}$, the group G can be defined as the $K(x)$-subgroup scheme of $GL(\mathcal{M}_{K(x)})$ that fixes a line L' in $W^* \otimes W$, i.e., such that $\Sigma_q^{\kappa_v}$ acts as the identity on $L' \otimes_\mathcal{A} \mathcal{A}/(\phi_v)$, for almost all $v \in \mathcal{C}$. It follows from Theorem 5.1 that the minimal submodule $\mathcal{W}'$ that contains L' becomes trivial over $K(x)$. Since $\mathcal{W}'_{K(x)}$ is contained in some construction of linear algebra of $\mathcal{M}_{K(x)}$, we have a functorial surjective group morphism

$$Gal(\mathcal{M}_{K(x)}) \longrightarrow Gal(\mathcal{W}'_{K(x)}) = \{1\}.$$

We conclude that $Gal(\mathcal{M}_{K(x)})$ acts trivially over $\mathcal{W}'_{K(x)}$, and therefore that $Gal(\mathcal{M}_{K(x)})$ is contained in G. □

NOTATION 6.12. To conclude, we point out that Theorem 6.8 is nothing but an equivalent geometric reformulation of Theorem 5.1. The core of this equivalence being Corollary 6.11, that allows to translate the triviality of the intrinsic Galois group in terms of the rationality of the solutions.

6.3. Finite intrinsic Galois groups

We deduce from Theorem 6.8 the following description of a finite intrinsic Galois group:

COROLLARY 6.13. *The following facts are equivalent:*

(1) *There exists a positive integer r such that the q-difference module $\mathcal{M} = (M, \Sigma_q)$ becomes trivial as a $\widetilde{q}$-difference module over $K(\widetilde{q}, t)$, with $\widetilde{q}^r = q$, $t^r = x$.*
(2) *There exists a positive integer r such that, for almost all $v \in \mathcal{C}$, the morphism $\Sigma_q^{\kappa_v r}$ induces the identity on $M \otimes_\mathcal{A} \mathcal{A}/(\phi_v)$.*
(3) *There exists a q-difference field extension $\mathcal{F}/K(x)$ of finite degree such that $\mathcal{M}$ becomes trivial over $\mathcal{F}$.*
(4) *The (intrinsic) Galois group of $\mathcal{M}$ is finite.*

In particular, if $Gal(\mathcal{M}_{K(x)})$ is finite, it is necessarily cyclic (of order r, if one chooses r minimal in the assertions above).

PROOF. The equivalence "1 ⇔ 2" follows from Theorem 5.1 applied to the $\widetilde{q}$-difference module $(M \otimes K(\widetilde{q}, t), \Sigma_q \otimes \sigma_{\widetilde{q}})$, over the field $K(\widetilde{q}, t)$.

If the intrinsic Galois group is finite, the reduction modulo ϕ_v of $\Sigma_q^{\kappa_v}$ must be a cyclic operator of order dividing the cardinality of $Gal(\mathcal{M}_{K(x)})$. So we have proved that "4 ⇒ 2". On the other hand, assertion 2 implies, by Theorem 6.8, that there exists a basis of $M_{K(x)}$ such that the representation of $Gal(\mathcal{M}_{K(x)})$ is given by the group of diagonal matrices, whose diagonal entries are r-th roots of unity.

Of course, assertion 1 implies assertion 3. The inverse implication follows from the Corollary 1.17, applied to a cyclic basis of $\mathcal{M}_{K(x)}$. □

6.4. Intrinsic Galois group of a q-difference module over $\mathbb{C}(x)$, for $q \neq 0, 1$

We deduce from the previous section a curvature characterization of the intrinsic Galois group of a q-difference module over $\mathbb{C}(x)$ for $q \in \mathbb{C} \smallsetminus \{0, 1\}$.[2]

Let $\mathcal{M}_{\mathbb{C}(x)} = (M_{\mathbb{C}(x)}, \Sigma_q)$ be a q-difference module over $\mathbb{C}(x)$. We can consider a finitely generated extension K of $\mathbb{Q}$ such that there exists a q-difference module $\mathcal{M}_{K(x)} = (M_{K(x)}, \Sigma_q)$ satisfying $\mathcal{M}_{\mathbb{C}(x)} = \mathcal{M}_{K(x)} \otimes_{K(x)} \mathbb{C}(x)$.

With an abuse of language, Theorem 5.1 can be rephrased as:

THEOREM 6.14. *The q-difference module $\mathcal{M}_{\mathbb{C}(x)} = (M_{\mathbb{C}(x)}, \Sigma_q)$ is trivial if and only if there exists a finitely generated extension K of $\mathbb{Q}$, a set of places $\mathcal{C}$ as in Chapter 5 and a q-difference module $\mathcal{M}_{K(x)}$ such that $\mathcal{M}_{\mathbb{C}(x)} \cong \mathcal{M}_{K(x)} \otimes_{K(x)} \mathbb{C}(x)$ and $\mathcal{M}_{K(x)}$ has zero v-curvature, for almost all $v \in \mathcal{C}$.*

We can of course define as in the previous sections an intrinsic Galois group $Gal(\mathcal{M}_{\mathbb{C}(x)})$. A noetherianity argument, that we have already used several times, shows the following:

PROPOSITION 6.15. *In the notation above we have:*
$$Gal(\mathcal{M}_{\mathbb{C}(x)}) \subset Gal(\mathcal{M}_{K(x)}) \otimes_{K(x)} \mathbb{C}(x).$$
Moreover there exists a finitely generated extension K' of K such that
$$Gal(\mathcal{M}_{K(x)} \otimes_{K(x)} K'(x)) \otimes_{K'(x)} \mathbb{C}(x) \cong Gal(\mathcal{M}_{\mathbb{C}(x)}).$$

Choosing K large enough, we can assume that $K = K'$, which we will do implicitly in the following informal statement. We can deduce from Theorem 6.14:

THEOREM 6.16. *The intrinsic Galois group $Gal(\mathcal{M}_{\mathbb{C}(x)})$ is the smallest $\mathbb{C}(x)$-subgroup scheme of $GL(M_{\mathbb{C}(x)})$ that contains the v-curvature of the q-difference module $\mathcal{M}_{K(x)}$, for K large enough and for almost all $v \in \mathcal{C}$.*

REMARK 6.17. By [**vdPS97**, Lemma 1.8], there exists a q-difference ring extension R of $\mathbb{C}(x)$, that is generated as $\mathbb{C}(x)$-algebra by a fundamental solution matrix of some q-difference system attached to $\mathcal{M}_{\mathbb{C}(x)}$ such that $R^{\sigma_q} = \mathbb{C}$ and R is σ_q-simple, that is, R has no non-trivial ideal setwise invariant by σ_q. The Galois group of R over $\mathbb{C}(x)$ is defined as the group of $\mathbb{C}(x)$-algebra automorphisms of

---

[2] All the statements in this subsection remain true if one replace $\mathbb{C}$ with any field of characteristic zero.

R, that commute with σ_q. It is the group of $\mathbb{C}$-points of a linear algebraic group $\mathcal{G}$ (see [**vdPS97**, Theorem 1.13]). In that framework, there is a complete Galois correspondence that allows to build a dictionary between the defining equations of the group $\mathcal{G}$ and the algebraic relations satisfied by the solutions of the q-difference system in R (see [**vdPS97**, Theorem 1.29]). One can deduce from Tannakian arguments that the intrinsic Galois group becomes isomorphic to $\mathcal{G}$ over a finite field extension of $\mathbb{C}(x)$ (see for instance [**And01**, §3.2.2.1]).

CHAPTER 7

The parametrized intrinsic Galois group

7.1. Differential and difference algebra

In this section, we briefly recall some notions of differential algebra as well as the geometric objects that one can define in this formalism. The interested reader can find an introduction to differential algebra and differential algebraic geometry in [**HSS16**] and a full presentation of this field in [**Kol73**]. We recall that all rings considered in this work are commutative with identity and contain the ring of integer numbers.

A differential ring (or ∂-ring for short) is a ring R together with a derivation $\partial : R \to R$, i.e., an additive map $\partial : R \to R$ satisfying the Leibniz rule $\partial(ab) = \partial(a)b + a\partial(b)$, for all $(a,b) \in R^2$. The ring of ∂-constants of R is $R^\partial = \{r \in R| \ \partial(r) = 0\}$. Any standard algebraic notion has a differential counterpart by requiring the compatibility of the algebraic structure with the derivation ∂: for instance a ∂-ideal $\mathfrak{q}$ of a ∂-ring R is an ideal of R that is setwise invariant by ∂, a ∂-field is a ∂-ring that is a field, etc. For k a ∂-field, the ∂-k-algebra $k\{x\}_\partial = k\{x_1, \ldots, x_n\}_\partial$ of ∂-polynomials over k in the ∂-variables $x_1, \ldots, x_n$ is the polynomial ring over k in the countable set of algebraically independent variables $x_1, \ldots, x_n, \partial(x_1), \ldots, \partial(x_n), \ldots$, with an action of ∂ as suggested by the names of the variables.

The notion of ∂-polynomials allows to develop a geometry where the varieties are defined as the zero sets of collections of ∂-polynomials. We won't give here a complete presentation of this geometry but we shall only focus on the notions that will be used in this paper. A ∂-field k is called differentially closed or ∂-closed, for short, if any system of ∂-polynomials with coefficients in k, having a solution in some differential field extension of k, has a solution in k. A differential closure of a ∂-field k is a ∂-field extension of k that is ∂-closed and can be k-embedded in any ∂-closed field extension of k.

DEFINITION 7.1. Let k be a ∂-field. A representable functor G from the category of ∂-k-algebras to the category of Groups is called *a ∂-k-group scheme*. The ∂-k-algebra representing G is denoted by $k\{G\}$ and called the ring of ∂-coordinates of G.

The Yoneda Lemma implies that the ring of ∂-coordinates of a ∂-k-group scheme is a ∂-k-Hopf algebra, that is a ∂-k-algebra that is an Hopf algebra whose structural maps commutes with ∂ (see [**Ovc09b**], §3.2 and 3.4). For instance, we denote by $\mathrm{GL}_\nu(k)$ the ∂-k-group scheme attached to the general linear group of size ν over k. It is represented by the ∂-k-algebra $k\left\{X, \frac{1}{\det(X)}\right\}_\partial$ where X is a $\nu \times \nu$ matrix of ∂-indeterminates. More generally, for any k-vector space V of finite dimension, we denote by $\mathbf{GL}(V)$ the ∂-k-group scheme of invertible k-linear automorphisms of V. As in the classical setting, one can define a ∂-k-subgroup scheme H of a ∂-k-group scheme as subfunctor of G. The ring of ∂-coordinates of H is the quotient of $k\{G\}$ by some ∂-ideal $\mathbb{I}(H)$, that we call defining ideal of H inside G.

It remains to explain some of the relations between the classical algebraic geometry and the differential algebraic geometry. Let V be a k-group scheme, i.e., a (covariant) functor from the category of k-algebras to the category of groups which is representable by a k- Hopf-algebra $k[V]$. We call $k[V]$ the ring of coordinates of V. In [**Gil02**], the author shows that the forgetful functor

$$\eta: \partial\text{-}k\text{-algebras} \to k\text{-algebras},$$

that associates to any ∂-k-algebra its underlying k-algebra, has a left adjoint denoted by D. This implies that the functor $\mathbf{V}$ from the category of ∂-k-algebras to the category of Groups, defined by the composition of V with the forgetful functor η is a ∂-k-group scheme, whose ring of ∂-coordinates is precisely $D(k[V])$. We call $\mathbf{V}$, the ∂-k-group scheme attached to V. The simple idea behind this construction is that polynomial equations are ∂-polynomials. More precisely if $V \subset \mathrm{GL}_\nu(k)$ and if $I(V) \subset k\left[X, \frac{1}{\det(X)}\right]$ is the vanishing ideal of V as subgroup scheme of $\mathrm{GL}_\nu(k)$ then the vanishing ideal of $\mathbf{V}$ as ∂-k-subgroup scheme of $\mathrm{GL}_\nu(k)$ is nothing else than the ∂-ideal generated by $I(V)$ in $k\left\{X, \frac{1}{\det(X)}\right\}_\partial$. Finally, Kolchin irreducibility theorem states that if $k[V]$ is a finitely generated integral k-algebra, then $D(k[V])$ is a finitely ∂-generated integral ∂-k-algebra and the dimension of V as k-scheme coincides with the ∂-dimension of $\mathbf{V}$ over k ([**Gil02**, §2]). Notice that we are calling $\mathrm{GL}_\nu(k)$ both the k-group scheme and the ∂-k-group scheme attached to the general linear group, anyway the context will always make clear to which one of the two structures we are referring to, without introducing complicate notation.

Conversely, given a ∂-k-subgroup scheme $\mathbf{V}$ of some $\mathrm{GL}_\nu(k)$, we can attach to $\mathbf{V}$ a k-subgroup scheme of $\mathrm{GL}_\nu(k)$ as follows. Let $\mathbb{I}(\mathbf{V}) \subset k\left\{X, \frac{1}{\det(X)}\right\}_\partial$ be the defining ideal of $\mathbf{V}$ in $\mathrm{GL}_\nu(k)$. Let $\mathbf{V}^Z$ be the k-subscheme of $\mathrm{GL}_\nu(k)$ defined by the ideal $\mathbb{I}(V) \cap k[X, \frac{1}{\det(X)}]$. We call $\mathbf{V}^Z$ the Zariski closure of $\mathbf{V}$ inside $\mathrm{GL}_\nu(k)$.

7.2. Parametrized intrinsic Galois groups

Let $\mathcal{F}$ be a σ_q-∂-field of *characteristic zero*, that is, an extension of $K(x)$ equipped with an extension of the q-difference operator σ_q and a derivation ∂ commuting to σ_q. For instance, the σ_q-∂-field $(K(x), \sigma_q, x\frac{d}{dx})$ satisfies these assumptions.

We can define an action of the derivation ∂ on the category $Diff(\mathcal{F}, \sigma_q)$, twisting the q-difference modules with the right $\mathcal{F}$-module $\mathcal{F}[\partial]_{\leq 1}$ of differential operators of order less than or equal to one. We recall that the structure of right $\mathcal{F}$-module on $\mathcal{F}[\partial]_{\leq 1}$ is defined via the Leibniz rule, i.e.,

$$\partial.\lambda = \lambda\partial + \partial(\lambda), \text{ for any } \lambda \in \mathcal{F}.$$

Let V be an $\mathcal{F}$-vector space. We denote by $F_\partial(V)$ the tensor product of the right $\mathcal{F}$-module $\mathcal{F}[\partial]_{\leq 1}$ with the left $\mathcal{F}$-module V:

$$F_\partial(V) := \mathcal{F}[\partial]_{\leq 1} \otimes_\mathcal{F} V.$$

We will write v for $1 \otimes v \in F_\partial(V)$ and $\partial(v)$ for $\partial \otimes v \in F_\partial(V)$, so that $av + b\partial(v) := (a + b\partial) \otimes v$, for any $v \in V$ and $a + b\partial \in \mathcal{F}[\partial]_{\leq 1}$. We endow $F_\partial(V)$ with a left $\mathcal{F}$-module structure such that if $\lambda \in \mathcal{F}$:

$$\lambda \partial(v) = \partial(\lambda v) - \partial(\lambda)v, \text{ for all } v \in V,$$

which means that $\lambda(\partial \otimes v) = \partial \otimes \lambda v - 1 \otimes \partial(\lambda)v$.

DEFINITION 7.2. The prolongation functor F_∂ is defined on the category of $\mathcal{F}$-vector spaces as follows. It associates to any object V the $\mathcal{F}$-vector space $F_\partial(V)$. If $f : V \to W$ is a morphism of $\mathcal{F}$-vector space then we define

$$F_\partial(f) : F_\partial(V) \to F_\partial(W),$$

setting $F_\partial(f)(\partial^i(v)) = \partial^i(f(v))$, for any $i = 0, 1$ and any $v \in V$ (using the convention that ∂^0 is the identity).

The prolongation functor F_∂ restricts to a functor from the category $Diff(\mathcal{F}, \sigma_q)$ to itself in the following way:

(1) If $\mathcal{M}_\mathcal{F} := (M_\mathcal{F}, \Sigma_q)$ is an object of $Diff(\mathcal{F}, \sigma_q)$ then $F_\partial(\mathcal{M}_\mathcal{F})$ is the q-difference module, whose underlying $\mathcal{F}$-vector space is $F_\partial(M_\mathcal{F}) = \mathcal{F}[\partial]_{\leq 1} \otimes M_\mathcal{F}$, as above, equipped with the q-invertible σ_q-semilinear operator defined by $\Sigma_q(\partial^i(m)) := \partial^i(\Sigma_q(m))$ for $i = 0, 1$.
(2) If $f \in Hom(\mathcal{M}_\mathcal{F}, \mathcal{N}_\mathcal{F})$ then $F_\partial(f)$ is defined in the same way as for $\mathcal{F}$-vector spaces.

REMARK 7.3. This formal definition comes from a simple and concrete idea. Let $\mathcal{M}_\mathcal{F}$ be an object of $Diff(\mathcal{F}, \sigma_q)$. We fix a basis $\underline{e}$ of $\mathcal{M}_\mathcal{F}$ over $\mathcal{F}$ such that $\Sigma_q \underline{e} = \underline{e} A$. Then $(\underline{e}, \partial(\underline{e}))$ is a basis of $F_\partial(\mathcal{M}_\mathcal{F})$ and

$$\Sigma_q(\underline{e}, \partial(\underline{e})) = (\underline{e}, \partial(\underline{e})) \begin{pmatrix} A & \partial A \\ 0 & A \end{pmatrix}.$$

In other terms, if $\sigma_q(Y) = A^{-1}Y$ is a q-difference system associated to $\mathcal{M}_\mathcal{F}$ with respect to a fixed basis $\underline{e}$, the q-difference system associated to $F_\partial(\mathcal{M}_\mathcal{F})$ with respect to the basis $\underline{e}, \partial(\underline{e})$ is:

$$\sigma_q(Z) = \begin{pmatrix} A^{-1} & \partial(A^{-1}) \\ 0 & A^{-1} \end{pmatrix} Z = \begin{pmatrix} A & \partial A \\ 0 & A \end{pmatrix}^{-1} Z.$$

If Y is a solution of $\sigma_q(Y) = A^{-1}Y$ in some σ_q-∂-extension of $\mathcal{F}$ then we have:

$$\sigma_q \begin{pmatrix} \partial Y & Y \\ Y & 0 \end{pmatrix} = \begin{pmatrix} A^{-1} & \partial(A^{-1}) \\ 0 & A^{-1} \end{pmatrix} \begin{pmatrix} \partial Y & Y \\ Y & 0 \end{pmatrix},$$

in fact the commutation of σ_q and ∂ implies:

$$\sigma_q(\partial Y) = \partial(\sigma_q Y) = \partial(A^{-1} Y) = A^{-1} \partial Y + \partial(A^{-1}) Y.$$

Let V be a finite dimensional $\mathcal{F}$-vector space. We denote by $Constr_\mathcal{F}^\partial(V)$ the smallest family of finite dimensional $\mathcal{F}$-vector spaces containing V and closed with respect to the constructions of linear algebra (i.e., direct sums, tensor products, symmetric and antisymmetric products, duals. See §1.1.1.) and the functor F_∂. We will say that an element $Constr_\mathcal{F}^\partial(V)$ is a construction of differential algebra of V. By functoriality, the ∂-$\mathcal{F}$-group scheme[1] $\mathbf{GL}(V)$ operates on $Constr_\mathcal{F}^\partial(V)$. For example $g \in \mathbf{GL}(V)$ acts on $F_\partial(V)$ through $g(\partial^i(v)) = \partial^i(g(v))$, for $i = 0, 1$.

---

[1] We denote by $\mathbf{GL}(V)$ is the ∂-$\mathcal{F}$-group scheme attached to $GL(V)$ as in §7.1.

If we start with a q-difference module $\mathcal{M}_\mathcal{F} = (M_\mathcal{F}, \Sigma_q)$ over $\mathcal{F}$, then every object of $Constr^\partial_\mathcal{F}(M_\mathcal{F})$ has a natural structure of q-difference module (see also §1.1.1). We will denote $Constr^\partial_\mathcal{F}(\mathcal{M}_\mathcal{F})$ the family of q-difference modules obtained in this way.

DEFINITION 7.4. We call parametrized intrinsic Galois group of an object $\mathcal{M}_\mathcal{F} = (M_\mathcal{F}, \Sigma_q)$ of $Diff(\mathcal{F}, \sigma_q)$ the group defined by

$$Gal^\partial(\mathcal{M}_\mathcal{F}) := \Big\{ g \in \mathbf{GL}(M_\mathcal{F}) : g(N_\mathcal{F}) \subset N_\mathcal{F} \text{ for all } q\text{-difference submodule}$$
$$\mathcal{N}_\mathcal{F} = (N_\mathcal{F}, \Sigma_q) \text{ contained in an object of } Constr^\partial_\mathcal{F}(\mathcal{M}_\mathcal{F}) \Big\} \subset \mathbf{GL}(M_\mathcal{F}).$$

Similarly to §6, one has to understand the definition above in a functorial sense. More precisely, $Gal^\partial(\mathcal{M}_\mathcal{F})$ is a functor from the category of ∂-$\mathcal{F}$-algebras to the category of groups, that associates to any ∂-$\mathcal{F}$-algebra S, the subgroup of $\mathrm{GL}(M_\mathcal{F} \otimes S)$ that stabilizes $N_\mathcal{F} \otimes S$, for all the q-difference submodules $\mathcal{N}_\mathcal{F}$ over $\mathcal{F}$ of any object in $Constr^\partial_\mathcal{F}(\mathcal{M}_\mathcal{F})$, after scalar extension to S.

The proposition below shows that this functor is representable and thus defines a ∂-$\mathcal{F}$-group scheme.

PROPOSITION 7.5. *The group $Gal^\partial(\mathcal{M}_\mathcal{F})$ is a reduced ∂-$\mathcal{F}$-subgroup scheme of $\mathbf{GL}(M_\mathcal{F})$.*

PROOF. The proof is a differential analogue of [**And01**, §3.2.2.2]. □

REMARK 7.6. The parametrized intrinsic Galois group has also a Tannakian interpretation. In the formalism of differential Tannakian categories (see [**Ovc09a**]), the parametrized Galois group is the group of differential tensor automorphisms of the forgetful functor of the full differential Tannakian subcategory generated by $\mathcal{M}_\mathcal{F}$ in $Diff(\mathcal{F}, \sigma_q)$.

For further reference, we recall (a particular case of) the Ritt-Raudenbush theorem (*cf.* [**Kap57**, Theorem 7.1]):

THEOREM 7.7. *Let $(\mathcal{F}, \partial)$ be a differential field of characteristic zero. If R is a reduced differentially finitely generated ∂-$\mathcal{F}$-algebra then R is ∂-noetherian.*

This means that any ascending chain of *radical* ∂-ideals (i.e., radical ∂-stable ideals) is stationary or equivalently that every radical ∂-ideal has a finite set of generators as radical ∂-ideal (which in general does not mean that it is a finitely generated ∂-ideal). Theorem 7.7 combined with Proposition 7.5 asserts that the parametrized intrinsic Galois group as well as any $\mathbf{GL}(\mathcal{F}^\nu)$ are ∂-noetherian.

The ∂-noetherianity of $\mathbf{GL}(\mathcal{F}^\nu)$ implies the following:

COROLLARY 7.8. *The parametrized intrinsic Galois group $Gal^\partial(\mathcal{M}_\mathcal{F})$ can be defined as the stabilizer of a line in a construction of differential algebra of $\mathcal{M}_\mathcal{F}$. This line can be chosen so that it is also a q-difference submodule of some construction of differential algebra of $\mathcal{M}_\mathcal{F}$.*

PROOF. Since $\mathbf{GL}(M_\mathcal{F})$ is ∂-noetherian, any descending chain of reduced differential subschemes in $\mathbf{GL}(M_\mathcal{F})$ is stationary. Then, let $\{\mathcal{W}^{(i)}; i \in I_h\}_h$ be an ascending chain of finite sets of q-difference submodules contained in some elements of $Constr^\partial(\mathcal{M}_{K(x)})$ so that all q-difference submodules contained in a construction of differential algebra are contained in some $\{\mathcal{W}^{(i)}; i \in I_h\}$. Let $\mathcal{G}_h$ be the

∂-$\mathcal{F}$-subgroup scheme of $\mathbf{GL}(M_\mathcal{F})$ defined as the stabilizer of $\{\mathcal{W}^{(i)}; i \in I_h\}$. By Cartier's theorem [**Wat79**, §11.4], the $\mathcal{G}_h$ are reduced. Then, the descending chain of ∂-$\mathcal{F}$-subgroup schemes $\mathcal{G}_h$ of $\mathbf{GL}(M_\mathcal{F})$ is stationary (see previous proposition). This proves that $Gal^\partial(\mathcal{M}_\mathcal{F})$ is the stabilizer of a finite number of q-difference submodules $\mathcal{W}^{(i)}$, $i \in I$, contained in some elements of $Constr^\partial(\mathcal{M}_{K(x)})$. It follows from a standard argument of linear algebra that $Gal^\partial(\mathcal{M}_\mathcal{F})$ is the stabilizer of the maximal exterior power of the direct sum of the $\mathcal{W}^{(i)}$'s (see Remark 6.3). □

Let $Gal(\mathcal{M}_\mathcal{F})$ be the intrinsic Galois group defined in the previous chapter and let $\mathbf{Gal}(\mathcal{M}_\mathcal{F})$ its associated ∂-$\mathcal{F}$-group scheme, defined as in §7.1. We have the following inclusion, that we will characterize in a more precise way in the next pages:

LEMMA 7.9. *Let $\mathcal{M}_\mathcal{F}$ be an object of $Diff(\mathcal{F}, \sigma_q)$. The following inclusion of ∂-$\mathcal{F}$-group schemes holds*

$$Gal^\partial(\mathcal{M}_\mathcal{F}) \subset \mathbf{Gal}(\mathcal{M}_\mathcal{F}).$$

REMARK 7.10. The inclusion above means that, for all ∂-$\mathcal{F}$-algebra S, we have $Gal^\partial(\mathcal{M}_\mathcal{F})(S) \subset \mathbf{Gal}(\mathcal{M}_\mathcal{F})(\mathbf{S})$. In particular, the parametrized intrinsic Galois group is a ∂-$\mathcal{F}$-subgroup scheme of the intrinsic Galois group, viewed as a ∂-$\mathcal{F}$-group scheme (see §7.1). Later, for $\mathcal{F} = K(x)$, we will prove that $Gal^\partial(\mathcal{M}_\mathcal{F})$ is actually Zariski dense in $\mathbf{Gal}(\mathcal{M}_\mathcal{F})$.

PROOF. We recall, that the $\mathcal{F}$-group scheme $Gal(\mathcal{M}_\mathcal{F})$ is defined as the stabilizer in $\mathrm{GL}(M_\mathcal{F})$ of all the subobjects contained in a construction of linear algebra of $\mathcal{M}_\mathcal{F}$. Because the list of subobjects contained in a construction of differential algebra of $\mathcal{M}_\mathcal{F}$ includes those contained in a construction of linear algebra of $\mathcal{M}_\mathcal{F}$, we get the claimed inclusion. □

7.3. Characterization of the parametrized intrinsic Galois group by curvatures

From now on we focus on the special case $\mathcal{F} = K(x)$, where K is a finitely generated extension of $\mathbb{Q}$. We endow $K(x)$ with the derivation $\partial := x\frac{d}{dx}$, that commutes with σ_q. We refer to Chapter 5 for notation.

Let $\mathcal{M}_{K(x)} = (M_{K(x)}, \Sigma_q)$ be a q-difference module. The differential version of Chevalley's theorem (*cf.* [**Cas72**, Proposition 14], [**MO11**, Theorem 5.1]) implies that any ∂-$K(x)$-subgroup scheme G of $\mathbf{GL}(M_{K(x)})$ can be defined as the stabilizer of some line $L_{K(x)}$ contained in a construction of differential algebra $\mathcal{W}_{K(x)}$ of $\mathcal{M}_{K(x)}$. Let M be a Σ_q-stable $\mathcal{A}$-lattice of $M_{K(x)}$ for some q-difference algebra $\mathcal{A}$. Up to enlarging $\mathcal{A}$, one finds an $\mathcal{A}$-lattice L of $L_{K(x)}$ and an $\mathcal{A}$-lattice W of $W_{K(x)}$. The latter is the underlying space of a q-difference module $\mathcal{W} = (W, \Sigma_q)$ over $\mathcal{A}$.

DEFINITION 7.11. Let $\widetilde{\mathcal{C}}$ be a non-empty cofinite subset of $\mathcal{C}$ and $(\Lambda_v)_{v \in \widetilde{\mathcal{C}}}$ be a family of invertible $\mathcal{A}/(\phi_v)$-linear operators acting on $M \otimes_\mathcal{A} \mathcal{A}/(\phi_v)$, respectively, for any $v \in \widetilde{\mathcal{C}}$. We say that a ∂-$K(x)$-group scheme $G = Stab(L_{K(x)})$ over $K(x)$ contains the operators Λ_v modulo ϕ_v, for almost all $v \in \mathcal{C}$, if for almost all (i.e. for almost all and at least one) $v \in \widetilde{\mathcal{C}}$ the operator Λ_v stabilizes $L \otimes_\mathcal{A} \mathcal{A}/(\phi_v)$ inside $W \otimes_\mathcal{A} \mathcal{A}/(\phi_v)$:

$$\Lambda_v \in Stab_{\mathcal{A}/(\phi_v)}(L \otimes_\mathcal{A} \mathcal{A}/(\phi_v)).$$

REMARK 7.12. The differential Chevalley's theorem and the ∂-noetherianity of $\mathrm{GL}(\mathcal{M}_{K(x)})$ imply that the notions of a ∂-$K(x)$-group scheme containing the operators Λ_v modulo ϕ_v, for almost all $v \in \mathcal{C}$, and the smallest ∂-$K(x)$-subgroup scheme of $\mathbf{GL}(\mathcal{M}_{K(x)})$ containing the operators Λ_v modulo ϕ_v, for almost all $v \in \mathcal{C}$, are well-defined. In particular they are independent of the choice of $\mathcal{A}$, $\mathcal{M}$ and $L_{K(x)}$ (See [**DV02**, 10.1.2] and Remark 6.6).

The main result of this section is the following:

THEOREM 7.13. *The ∂-$K(x)$-group scheme $Gal^{\partial}(\mathcal{M}_{K(x)})$ is the smallest ∂-$K(x)$-subgroup scheme of $\mathbf{GL}(M_{K(x)})$ that contains the operators $\Sigma_q^{\kappa_v}$ modulo ϕ_v, for almost all $v \in \mathcal{C}$.*

PROOF. The lemmas below plus the differential Chevalley theorem allow to prove Theorem 7.13 in exactly the same way as Theorem 6.8. □

LEMMA 7.14. *The ∂-$K(x)$-group scheme $Gal^{\partial}(\mathcal{M}_{K(x)})$ contains the operators $\Sigma_q^{\kappa_v}$ modulo ϕ_v, for almost all $v \in \mathcal{C}$.*

PROOF. The statement follows immediately from the fact that $Gal^{\partial}(\mathcal{M}_{K(x)})$ can be defined as the stabilizer of a rank one q-difference submodule of some construction of differential algebra of $\mathcal{M}_{K(x)}$, which is *a fortiori* stable under the action of $\Sigma_q^{\kappa_v}$. □

LEMMA 7.15. *$Gal^{\partial}(\mathcal{M}_{K(x)}) = \{1\}$ if and only if $\mathcal{M}_{K(x)}$ is a trivial q-difference module.*

PROOF. The proof is a differential Tannakian analogue of [**DV02**, Lemma 9.34]. See [**Ovc09a**] for the notions of differential Tannakian categories. □

We obtain the following:

COROLLARY 7.16. *The Zariski closure of the parametrized intrinsic Galois group $Gal^{\partial}(\mathcal{M}_{K(x)})$ coincides the algebraic intrinsic Galois group $Gal(\mathcal{M}_{K(x)})$.(see §7.1)*

PROOF. We have seen in Lemma 7.9 that $Gal^{\partial}(\mathcal{M}_{K(x)})$ is a subgroup of $\mathbf{Gal}(\mathcal{M}_{K(x)})$. By Theorem 7.13 (resp. Theorem 6.8) we have that the intrinsic Galois group $Gal^{\partial}(\mathcal{M}_{K(x)})$ (resp. $Gal(\mathcal{M}_{K(x)})$) is the smallest ∂-$K(x)$-subgroup scheme (resp.$K(x)$-subgroup scheme) of $\mathrm{GL}(M_{K(x)})$ that contains the operators $\Sigma_q^{\kappa_v}$ modulo ϕ_v, for almost all $v \in \mathcal{C}$. This immediately implies the Zariski density. □

EXAMPLE 7.17. The logarithm is solution both a q-difference and a differential system:

$$Y(qx) = \begin{pmatrix} 1 & \log q \\ 0 & 1 \end{pmatrix} Y(x), \quad \partial Y(x) = \begin{pmatrix} 0 & 1 \\ 0 & 0 \end{pmatrix} Y(x).$$

It is easy to verify that the two systems are compatible in the sense that $\partial \sigma_q Y(x) = \sigma_q \partial Y(x)$ (and therefore that the induced condition on the matrices of the systems is verified).

By iterating the q-difference system for any $n \in \mathbb{Z}_{>0}$ we obtain:

$$Y(q^n x) = \begin{pmatrix} 1 & n \log q \\ 0 & 1 \end{pmatrix} Y(x).$$

This implies that the parametrized intrinsic Galois group is the subgroup of $\mathbb{G}_{a,K(x)}$, the additive ∂-$K(x)$-group scheme, defined by the equation $\partial y = 0$.

7.4. Parametrized intrinsic Galois group of a q-difference module over $\mathbb{C}(x)$, for $q \neq 0, 1$

We conclude with some remarks on complex q-difference modules. Let $\mathcal{M}_{\mathbb{C}(x)} = (M_{\mathbb{C}(x)}, \Sigma_q)$ be a q-difference module over $\mathbb{C}(x)$. We can consider a finitely generated extension of K of $\mathbb{Q}$ such that there exists a q-difference module $\mathcal{M}_{K(x)} = (M_{K(x)}, \Sigma_q)$ satisfying $\mathcal{M}_{\mathbb{C}(x)} = \mathcal{M}_{K(x)} \otimes_{K(x)} \mathbb{C}(x)$. We can of course define, as above, two parametrized intrinsic Galois groups, $Gal^\partial(\mathcal{M}_{K(x)})$ and $Gal^\partial(\mathcal{M}_{\mathbb{C}(x)})$. A (differential) noetherianity argument, that we have already used several times, on the submodules stabilized by those groups shows the following:

PROPOSITION 7.18. *In the notations above, we have:*
$$Gal^\partial(\mathcal{M}_{\mathbb{C}(x)}) \subset Gal^\partial(\mathcal{M}_{K(x)}) \otimes_{K(x)} \mathbb{C}(x).$$
Moreover there exists a finitely generated extension K' of K such that
$$Gal^\partial(\mathcal{M}_{K(x)} \otimes_{K(x)} K'(x)) \otimes_{K'(x)} \mathbb{C}(x) \cong Gal^\partial(\mathcal{M}_{\mathbb{C}(x)}).$$

We can informally rephrase Theorem 7.13 in the following way:

THEOREM 7.19. *The parametrized intrinsic Galois group $Gal^\partial(\mathcal{M}_{\mathbb{C}(x)})$ is the smallest ∂-$\mathbb{C}(x)$-subgroup scheme of $\mathbf{GL}(M_{\mathbb{C}(x)})$ that contains a non-empty cofinite set of curvatures of the q-difference module $\mathcal{M}_{K'(x)}$.*

REMARK 7.20. In [**HS08**], the authors develop a parametrized Galois theory for q-difference systems over $\widetilde{\mathbb{C}}(x)$ where $\widetilde{\mathbb{C}}$ is a differential closure of $\mathbb{C}$. This theory generalizes the Picard-Vessiot theory of [**vdPS97**]. The parametrized Galois theory allows to encode the differential algebraic relations satisfied by the solutions of a q-difference system in the geometric structure of a parametrized Galois group. The parametrized Galois group is a ∂-$\widetilde{\mathbb{C}}$-group scheme and has a differential Tannakian interpretation (see for instance [**GGO13**, §5.3]). Combining arguments of [**And01**, §3.2.2.1] and [**GGO13**, Prop. 4.28 and 4.29] one can prove that the parametrized intrinsic Galois group of a q-difference system over $\mathbb{C}(x)$ becomes isomorphic to the parametrized Galois group of [**HS08**] over a differential closure of $\widetilde{\mathbb{C}}(x)$.

7.5. The example of the Jacobi Theta function

Consider the Jacobi Theta function
$$\Theta(x) = \sum_{n \in \mathbb{Z}} q^{-n(n-1)/2} x^n,$$
which is solution of the q-difference equation
$$\Theta(qx) = qx\Theta(x).$$
Iterating the equation, one proves that Θ satisfies $y(q^n x) = q^{n(n+1)/2} x^n y(x)$, for any $n \geq 0$. Therefore we immediately deduce that the intrinsic Galois group of the rank one q-difference module $\mathcal{M}_\Theta = (K(x).\Theta, \Sigma_q)$, with
$$\begin{aligned}\Sigma_q : K(x).\Theta &\longrightarrow K(x).\Theta \\ f(x)\Theta &\longmapsto f(qx)qx\Theta\end{aligned},$$

is the whole multiplicative group $\mathbb{G}_{m,K(x)}$. As far as the parametrized intrinsic Galois group is concerned we have:

PROPOSITION 7.21. *The parametrized intrinsic Galois group $Gal^\partial(\mathcal{M}_\Theta)$ is defined by $\partial(\partial(y)/y) = 0$.*

PROOF. For almost any $v \in \mathcal{C}$, the reduction modulo ϕ_v of $q^{\kappa_v(\kappa_v+1)/2} x^{\kappa_v}$ is the monomial x^{κ_v}, which satisfies the equation $\partial\left(\frac{\partial x^{\kappa_v}}{x^{\kappa_v}}\right) = 0$. This means that parametrized intrinsic Galois group $Gal^\partial(\mathcal{M}_\Theta)$ is a subgroup of the ∂-$K(x)$-subgroup scheme defined by $\partial\left(\frac{\partial y}{y}\right) = 0$. In other words, the logarithmic derivative

$$\begin{array}{rcl} \mathbb{G}_m & \longrightarrow & \mathbb{G}_a \\ y & \longmapsto & \frac{\partial y}{y} \end{array}$$

sends $Gal^\partial(\mathcal{M}_\Theta, \eta_{K(x)})$ into a subgroup of the additive group $\mathbb{G}_{a,K(x)}$ defined by the equation $\partial z = 0$. This is nothing else that $\mathbb{G}_{a,K}$, whose proper K-subgroup scheme is only $\{0\}$. If the image by the logarithmic derivative of $Gal^\partial(\mathcal{M}_\Theta)$ were $\{0\}$, then the curvatures should be constant with respect to ∂. It is not the case, which ends the proof. □

Let us consider a norm $|\ |$ on K such that $|q| \neq 1$. The differential dimension of the subgroup $\partial\left(\frac{\partial y}{y}\right) = 0$ is zero. Using the comparison of the parametrized Galois group of [**HS08**] and the parametrized intrinsic Galois group, we obtain that Θ is differentially algebraic over the field of rational functions $\widetilde{C}_E(x)$ with coefficients in the differential closure $\widetilde{C}_E$ of the elliptic function over $K^*/q^\mathbb{Z}$. In fact, the function Θ satisfies

$$\sigma_q\left(\frac{\partial \Theta}{\Theta}\right) = \frac{\partial \Theta}{\Theta} + 1,$$

which implies that $\partial\left(\frac{\partial \Theta}{\Theta}\right)$ is an elliptic function. Since the Weierstrass function is differentially algebraic over $K(x)$, the Jacobi Theta function is also differentially algebraic over $K(x)$.

Notice that, if q is transcendental over $\mathbb{Q}$, the derivation $\frac{d}{dq}$ naturally comes into the picture. Since it intertwines with σ_q in a relatively complicate way, the study of this situation requires a specific approach. See [**DVH12**].

Part 4

Comparison with the non-linear theory

CHAPTER 8

Preface to Part 4. The Galois D-groupoid of a q-difference system, by Anne Granier

We recall here the definition of the Galois D-groupoid of a q-difference system, and how to recover groups from it in the case of a linear q-difference system. This preface thus consists in a summary of Chapter 3 of [**Gra09**].

8.1. Definitions

We need to recall first Malgrange's definition of D-groupoids, following [**Mal01**] but specializing it to the base space $\mathbb{P}^1_\mathbb{C} \times \mathbb{C}^\nu$ as in [**Gra09**] and [**Gra12**], and to explain how it allows to define a Galois D-groupoid for q-difference systems.

Fix $\nu \in \mathbb{N}^*$, and denote by M the analytic complex variety $\mathbb{P}^1_\mathbb{C} \times \mathbb{C}^\nu$. We call *local diffeomorphism of M* any biholomorphism between two open sets of M, and we denote by $Aut(M)$ the set of germs of local diffeomorphisms of M. Essentially, a D-groupoid is a subgroupoid of $Aut(M)$ defined by a system of partial differential equations.

Let us precise what is the object which represents the system of partial differential equations in this rough definition.

A germ of a local diffeomorphism of M is determined by the coordinates denoted by $(x, X) = (x, X_1, \ldots, X_\nu)$ of its source point, the coordinates denoted by $(\bar{x}, \bar{X}) = (\bar{x}, \bar{X}_1, \ldots, \bar{X}_\nu)$ of its target point, and the coordinates denoted by $\frac{\partial \bar{x}}{\partial x}, \frac{\partial \bar{x}}{\partial X_1}, \ldots, \frac{\partial \bar{X}_1}{\partial x}, \ldots, \frac{\partial^2 \bar{x}}{\partial x^2}, \ldots$ which represent its partial derivatives evaluated at the source point. We also denote by δ the polynomial in the coordinates above, which represents the Jacobian of a germ evaluated at the source point. We will allow ourselves to use abbreviations for some sets of these coordinates, as for example $\frac{\partial \bar{X}}{\partial X}$ to represent all the coordinates $\frac{\partial \bar{X}_i}{\partial X_j}$ and $\partial \bar{X}$ for all the coordinates $\frac{\partial \bar{X}_i}{\partial x_j}$, $\frac{\partial \bar{X}_i}{\partial \bar{x}_j}, \frac{\partial \bar{X}_i}{\partial X_j}$ and $\frac{\partial \bar{X}_i}{\partial \bar{X}_j}$.

We denote by r any positive integer. We call *partial differential equation*, or only *equation*, of order $\leq r$ any function $E(x, X, \bar{x}, \bar{X}, \partial \bar{x}, \partial \bar{X}, \ldots, \partial^r \bar{x}, \partial^r \bar{X})$ which locally and holomorphically depends on the source and target coordinates, and polynomially on δ^{-1} and on the partial derivative coordinates of order $\leq r$. These equations are endowed with a sheaf structure on $M \times M$ which we denote by $\mathcal{O}_{J^*_r(M,M)}$. We then denote by $\mathcal{O}_{J^*(M,M)}$ the sheaf of all the equations, that is the direct limit of the sheaves $\mathcal{O}_{J^*_r(M,M)}$. It is endowed with natural derivations of the equations with respect to the source coordinates. For example, one has: $D_x \cdot \frac{\partial \bar{X}_i}{\partial X_j} = \frac{\partial^2 \bar{X}_i}{\partial x \partial X_j}$.

To formulate the definition of D-groupoid, we will consider a pseudo-coherent (in the sense of [**Mal01**]) differential ideal[1] $\mathcal{I}$ of $\mathcal{O}_{J^*(M,M)}$. A *solution* of such

---

[1] We will say everywhere differential ideal for sheaf of differential ideal.

an ideal $\mathcal{I}$ is a germ of a local diffeomorphism $g : (M,a) \to (M,g(a))$ such that, for any equation E of the fiber $\mathcal{I}_{(a,g(a))}$, the function defined by $(x,X) \mapsto E((x,X), g(x,X), \partial g(x,X), \ldots)$ is null in a neighborhood of a in M. The solutions of $\mathcal{I}$ is denoted by $sol(\mathcal{I})$ and forms a set groupoid.

The set $Aut(M)$ is endowed with a groupoid structure for the composition c and the inversion i of the germs of local diffeomorphisms of M. We thus have to characterize, with the comorphisms c^* and i^* defined on $\mathcal{O}_{J^*(M,M)}$, the systems of partial differential equations $\mathcal{I} \subset \mathcal{O}_{J^*(M,M)}$ whose set of solutions $sol(\mathcal{I})$ is a subgroupoid of $Aut(M)$.

We call *groupoid of order r* on M the subvariety of the space of invertible jets of order r defined by a coherent ideal $\mathcal{I}_r \subset \mathcal{O}_{J^*_r(M,M)}$ such that:

 (i) all the germs of the identity map of M are solutions of $\mathcal{I}_r$,
 (ii) $c^*(\mathcal{I}_r) \subset \mathcal{I}_r \otimes \mathcal{O}_{J^*_r(M,M)} + \mathcal{O}_{J^*_r(M,M)} \otimes \mathcal{I}_r$,
 (iii) $\iota^*(\mathcal{I}_r) \subset \mathcal{I}_r$.

The solutions of such an ideal $\mathcal{I}_r$ form a subgroupoid of $Aut(M)$.

DEFINITION 8.1. According to [**Mal01**], a *D-groupoid* $\mathcal{G}$ on M is a subvariety of the space $(M^2, \mathcal{O}_{J^*(M,M)})$ of invertible jets defined by a reduced, pseudo-coherent differential ideal $\mathcal{I}_\mathcal{G} \subset \mathcal{O}_{J^*(M,M)}$ such that

 (i') all the germs of the identity map of M are solutions of $\mathcal{I}_\mathcal{G}$,
 (ii') for any relatively compact open set U of M, there exists a closed complex analytic subvariety Z of U of codimension ≥ 1, and a positive integer $r_0 \in \mathbb{N}$ such that, for all $r \geq r_0$ and denoting by $\mathcal{I}_{\mathcal{G},r} = \mathcal{I}_\mathcal{G} \cap \mathcal{O}_{J^*_r(M,M)}$, one has, above $(U \setminus Z)^2$: $c^*(\mathcal{I}_{\mathcal{G},r}) \subset \mathcal{I}_{\mathcal{G},r} \otimes \mathcal{O}_{J^*_r(M,M)} + \mathcal{O}_{J^*_r(M,M)} \otimes \mathcal{I}_{\mathcal{G},r}$,
 (iii') $\iota^*(\mathcal{I}_\mathcal{G}) \subset \mathcal{I}_\mathcal{G}$.

The ideal $\mathcal{I}_\mathcal{G}$ totally determines the D-groupoid $\mathcal{G}$, so we will rather focus on the ideal $\mathcal{I}_\mathcal{G}$ than its solution $sol(\mathcal{I}_\mathcal{G})$ in $Aut(M)$. Thanks to the analytic continuation theorem, $sol(\mathcal{I}_\mathcal{G})$ is a subgroupoid of $Aut(M)$.

The flexibility introduced by Malgrange in his definition of D-groupoid allows him to obtain two main results. Theorem 4.4.1 of [**Mal01**] states that the reduced differential ideal of $\mathcal{O}_{J^*(M,M)}$ generated by a coherent ideal $\mathcal{I}_r \subset \mathcal{O}_{J^*_r(M,M)}$ which satisfies the previous conditions *(i),(ii)*, and *(iii)* defines a D-groupoid on M. Theorem 4.5.1 of [**Mal01**] states that for any family of D-groupoids on M defined by a family of ideals $\{\mathcal{G}^i\}_{i \in I}$, the ideal $\sqrt{\sum \mathcal{G}^i}$ defines a D-groupoid on M called *intersection*. The terminology is legitimated by the equality: $sol(\sqrt{\sum \mathcal{G}^i}) = \cap_{i \in I} sol(\mathcal{G}^i)$. This last result allows to define the notion of D-envelope of any subgroupoid of $Aut(M)$.

Fix $q \in \mathbb{C}^*$, and let $Y(qx) = F(x, Y(x))$ be a (non-linear) q-difference system, with $F(x,X) \in \mathbb{C}(x,X)^\nu$. Consider the set subgroupoid of $Aut(M)$ generated by the germs of the application $(x,X) \mapsto (qx, F(x,X))$ at any point of M where it is well-defined and invertible, and denote it by $Dyn(F)$. The Galois D-groupoid of the q-difference system $Y(qx) = F(x,Y(x))$ is the D-enveloppe of $Dyn(F)$, that is the *intersection* of the D-groupoids on M whose set of solutions contains $Dyn(F)$.

8.2. A bound for the Galois D-groupoid of a linear q-difference system

For all the following, consider a rational linear q-difference system $Y(qx) = A(x)Y(x)$, with $A(x) \in GL_\nu(\mathbb{C}(x))$. We denote by $\mathcal{G}al(A(x))$ the Galois D-groupoid of this system as defined at the end of the previous section 8.1, we denote by $\mathcal{I}_{\mathcal{G}al(A(x))}$ its defining ideal of equations, and by $sol(\mathcal{G}al(A(x)))$ its groupoid of solutions.

The elements of the dynamics $Dyn(A(x))$ of $Y(qx) = A(x)Y(x)$ are the germs of the local diffeomorphisms of M of the form $(x, X) \mapsto (q^k x, A_k(x)X)$, with:

$$A_k(x) = \begin{cases} Id_n & \text{if } k = 0, \\ \prod_{i=0}^{k-1} A(q^i x) & \text{if } k \in \mathbb{N}^*, \\ \prod_{i=k}^{-1} A(q^i x)^{-1} & \text{if } k \in -\mathbb{N}^*. \end{cases}$$

The first component of these diffeomorphisms is independent on the variables X and depends linearly on the variable x, and the second component depends linearly on the variables X. These properties can be expressed in terms of partial differential equations. This gives an *upper bound* for the Galois D-groupoid $\mathcal{G}al(A(x))$ which is defined in the following proposition.

PROPOSITION 8.2. *The coherent ideal:*

$$\left\langle \frac{\partial \bar{x}}{\partial X}, \frac{\partial \bar{x}}{\partial x} x - \bar{x}, \partial^2 \bar{x}, \frac{\partial \bar{X}}{\partial X} X - \bar{X}, \frac{\partial^2 \bar{X}}{\partial X^2} \right\rangle \subset \mathcal{O}_{J_2^*(M,M)}$$

satisfies the conditions (i),(ii), and (iii) of 8.1. Hence, thanks to Theorem 4.4.1 of [**Mal01**], *the reduced differential ideal* $\mathcal{I}_{\mathcal{L}in}$ *it generates defines a D-groupoid* $\mathcal{L}in$. *Its solutions $sol(\mathcal{L}in)$ are the germs of the local diffeomorphisms of M of the form:*

$$(x, X) \mapsto (\alpha x, \beta(x)X),$$

with $\alpha \in \mathbb{C}^*$ and locally, $\beta(x) \in GL_\nu(\mathbb{C})$ for all x.

They contain $Dyn(A(x))$, and therefore, given the definition of $\mathcal{G}al(A(x))$, one has the inclusion

$$\mathcal{G}al(A(x)) \subset \mathcal{L}in,$$

which means that:

$$\mathcal{I}_{\mathcal{L}in} \subset \mathcal{I}_{\mathcal{G}al(A(x))} \quad and \quad sol(\mathcal{G}al(A(x))) \subset sol(\mathcal{L}in).$$

PROOF. *cf* proof of Proposition 3.2.1 of [**Gra09**] for more details. □

REMARK 8.3. Given their shape, the solutions of $\mathcal{L}in$ are naturally defined in neighborhoods of transversals $\{x_a\} \times \mathbb{C}^\nu$ of M. Actually, consider a particular element of $sol(\mathcal{L}in)$, that is precisely a germ at a point $(x_a, X_a) \in M$ of a local diffeomorphism g of M of the form $(x, X) \mapsto (\alpha x, \beta(x)X)$. Consider then a neighborhood Δ of x_a in $P^1\mathbb{C}$ where the matrix $\beta(x)$ is well-defined and invertible, consider the "cylinders" $T_s = \Delta \times \mathbb{C}^\nu$ and $T_t = \alpha\Delta \times \mathbb{C}^\nu$ of M, and the diffeomorphism $\tilde{g} : T_s \to T_t$ well-defined by $(x, X) \to (\alpha x, \beta(x)X)$. Therefore, according to the previous Proposition 8.2, all the germs of $\tilde{g}$ at the points of T_s are in $sol(\mathcal{L}in)$ too.

The defining ideal $\mathcal{I}_{\mathcal{L}in}$ of the bound $\mathcal{L}in$ is generated by very simple equations. This allows to reduce modulo $\mathcal{I}_{\mathcal{L}in}$ the equations of $\mathcal{I}_{\mathcal{G}al(A(x))}$ and obtain some simpler representative equations, in the sense that they only depend on some variables.

PROPOSITION 8.4. *Let $r \geq 2$. For any equation $E \in \mathcal{I}_{\mathcal{G}al(A(x))}$ of order r, there exists an invertible element $u \in \mathcal{O}_{J_r^*(M,M)}$, an equation $L \in \mathcal{I}_{\mathcal{L}in}$ of order r, and an* equation $E_1 \in \mathcal{I}_{\mathcal{G}al(A(x))}$ of order r only depending on the variables written below, such that:

$$uE = L + E_1\left(x, X, \frac{\partial \bar{x}}{\partial x}, \frac{\partial \bar{X}}{\partial X}, \frac{\partial^2 \bar{X}}{\partial x \partial X}, \cdots \frac{\partial^r \bar{X}}{\partial x^{r-1} \partial X}\right).$$

PROOF. The invertible element u is a power of δ. The proof consists then in performing the divisions of the equation uE, and then its successive remainders, by the generators of $\mathcal{I}_{\mathcal{L}in}$. More details are given in the proof of Proposition 3.2.3 of [**Gra09**]. □

8.3. Groups from the Galois D-groupoid of a linear q-difference system

We are going to prove that the solutions of the Galois D-groupoid $\mathcal{G}al(A(x))$ are, like the solutions of the bound $\mathcal{L}in$, naturally defined in neighborhoods of transversals of M. This property, together with the groupoid structure of $sol(\mathcal{G}al(A(x)))$, allows to exhibit groups from the solutions of $\mathcal{G}al(A(x))$ which fix the transversals.

According to Proposition 8.2, an element of $sol(\mathcal{G}al(A(x)))$ is also an element of $sol(\mathcal{L}in)$. Therefore, it is a germ at a point $a = (x_a, X_a) \in M$ of a local diffeomorphism $g : (M, a) \to (M, g(a))$ of the form $(x, X) \mapsto (\alpha x, \beta(x) X)$, such that, for any equation $E \in \mathcal{I}_{\mathcal{G}al(A(x))}$, one has $E((x, X), g(x, X), \partial g(x, X), \ldots) = 0$ in a neighborhood of a in M.

Consider an open connected neighborhood Δ of x_a in $\mathbb{P}^1_\mathbb{C}$ on which the matrix β is well-defined and invertible, that is where β can be prolongated in a matrix $\beta \in GL_\nu(\mathcal{O}(\Delta))$. Consider the "cylinders" $T_s = \Delta \times \mathbb{C}^\nu$ and $T_t = \alpha \Delta \times \mathbb{C}^\nu$ of M, and the diffeomorphism $\tilde{g} : T_s \to T_t$ defined by $(x, X) \to (\alpha x, \beta(x) X)$.

PROPOSITION 8.5. *The germs at all points of T_s of the diffeomorphism $\tilde{g}$ are elements of $sol(\mathcal{G}al(A(x)))$.*

PROOF. For all $r \in \mathbb{N}$, the ideal $(\mathcal{I}_{\mathcal{G}al(A(x))})_r = \mathcal{I}_{\mathcal{G}al(A(x))} \cap \mathcal{O}_{J_r^*(M,M)}$ is coherent. Thus, for any point $(y_0, \bar{y}_0) \in M^2$, there exists an open neighbourhood Ω of $(y_0, \bar{y}_0)$ in M^2, and equations $E_1^\Omega, \ldots, E_l^\Omega$ of $(\mathcal{I}_{\mathcal{G}al(A(x))})_r$ defined on the open set Ω such that:

$$\left((\mathcal{I}_{\mathcal{G}al(A(x))})_r\right)_{|\Omega} = \left(\mathcal{O}_{J_r^*(M,M)}\right)_{|\Omega} E_1^\Omega + \cdots + \left(\mathcal{O}_{J_r^*(M,M)}\right)_{|\Omega} E_l^\Omega.$$

Let $a_1 \in T_s = \Delta \times \mathbb{C}^\nu$. Let $\gamma : [0, 1] \to T_s$ be a path in T_s such that $\gamma(0) = a$ and $\gamma(1) = a_1$. Let $\{\Omega_0, \ldots, \Omega_N\}$ be a finite covering of the path $\gamma([0, 1]) \times \tilde{g}(\gamma([0, 1]))$ in $T_s \times T_t$ by connected open sets $\Omega_i \subset (T_s \times T_t)$ like above, and such that the origin $(\gamma(0), g(\gamma(0))) = (a, g(a))$ belongs to Ω_0.

The germ of g at the point a is an element of $sol(\mathcal{G}al(A(x)))$. Therefore, one has $E_k^{\Omega_0}((x, X), g(x, X), \partial g(x, X), \ldots) \equiv 0$ a neighbourhood of a, for all $1 \leq k \leq l$. Moreover, by analytic continuation, one has also $E_k^{\Omega_0}(x, X, \tilde{g}(x, X), \partial \tilde{g}(x, X), \ldots) \equiv 0$ on the source projection of Ω_0 in M. It means that the germs of $\tilde{g}$ at any point of the source projection of Ω_0 are solutions of $(\mathcal{I}_{\mathcal{G}al(A(x))})_r$.

Then, step by step, one gets that the germs of $\tilde{g}$ at any point of the source projection of Ω_k are solutions of $(\mathcal{I}_{\mathcal{G}al(A(x))})_r$ and, in particular, the germ of $\tilde{g}$ at the point a_1 is also a solution of $(\mathcal{I}_{\mathcal{G}al(A(x))})_r$. □

This Proposition 8.5 means that any solution of the Galois D-groupoid $\mathcal{G}al(A(x))$ is naturally defined in a neighbourhood of a transversal of M, above.

REMARK 8.6. In some sense, the "equations" counterpart of this proposition is Lemma 9.14.

The solutions of $\mathcal{G}al(A(x))$ which fix the transversals of M can be interpreted as solutions of a sub-D-groupoid of $\mathcal{G}al(A(x))$, partly because this property can be interpreted in terms of partial differential equations. Actually, a germ of a diffeomorphism of M fix the transversals of M if and only if it is a solution of the equation $\bar{x} - x$.

The ideal of $\mathcal{O}_{J_0^*(M,M)}$ generated by the equation $\bar{x} - x$ satisfies the conditions *(i)*,*(ii)*, and *(iii)* of 8.1. Hence, thanks to Theorem 4.4.1 of [**Mal01**], the reduced differential ideal it generates defines a D-groupoid:

DEFINITION 8.7. We call $\mathcal{T}rv$ the D-groupoid generated by the equation $\bar{x} - x$.

Its solutions, $sol(\mathcal{T}rv)$, are the germs of the local diffeomorphisms of M of the form: $(x, X) \mapsto (x, \bar{X}(x, X))$.

DEFINITION 8.8. We call $\widetilde{\mathcal{G}al(A(x))}$ the *intersection* D-groupoid $\mathcal{G}al(A(x)) \cap \mathcal{T}rv$, in the sense of Theorem 4.5.1 of [**Mal01**], whose defining ideal of equations $\mathcal{I}_{\widetilde{\mathcal{G}al(A(x))}}$ is generated by $\mathcal{I}_{\mathcal{G}al(A(x))}$ and $\mathcal{I}_{\mathcal{T}rv}$.

REMARK 8.9. Let $x_0 \in \mathbb{P}^1_{\mathbb{C}}$. Since by Proposition 8.2, the solutions of $\mathcal{G}al(A(x))$ are solutions of $\mathcal{L}in$, the solutions of $sol(\widetilde{\mathcal{G}al(A(x))})$ defined in a neighborhood of the transversal $\{x_0\} \times \mathbb{C}^\nu$ are of the form $(x, X) \mapsto (x, \beta(x)X)$ where $\beta(x) \in \mathrm{GL}_n(\mathbb{C}\{x - x_0\})$. The groupoid structure is given by composition of germs of diffeomorphism with compatible source and target points. Let $\phi_1 : (x, X) \mapsto (x, \beta_1(x)X)$ and $\phi_2 : (x, X) \mapsto (x, \beta_2(x)X)$ two solutions defined in a neighborhood of the transversal. In a neighborhood of $\{x_0\} \times \mathbb{C}^\nu$, where these germs are both defined, one can compose them and find a new germ of solution $\phi_2 \circ \phi_1$, that is given by $(x, X) \mapsto (x, \beta_2(x)\beta_1(x)X)$. The following proposition resume this discussion for which one can find a more detailed proof in [**Gra09**, Proposition 3.3.2].

PROPOSITION 8.10. Let $x_0 \in \mathbb{P}^1_{\mathbb{C}}$. The set of solutions of $\widetilde{\mathcal{G}al(A(x))}$ defined in a neighborhood of the transversal $\{x_0\} \times \mathbb{C}^\nu$ of M can be identified with a subgroup of $GL_\nu(\mathbb{C}\{x - x_0\})$.

In the particular case of a constant linear q-difference system, that is with $A(x) = A \in GL_\nu(\mathbb{C})$, the solutions of the Galois D-groupoid $\mathcal{G}al(A)$ are in fact global diffeomorphisms of M, and the set of those that fix the transversals of M can be identified with an algebraic subgroup of $GL_\nu(\mathbb{C})$. This can be shown using a better bound than $\mathcal{L}in$ for the Galois D-groupoid of a constant linear q-difference system (*cf* Proposition 3.4.2 of [**Gra09**]), or computing the D-groupoid $\mathcal{G}al(A)$ directly (*cf* Theorem 2.1 of [**Gra12**] or Theorem 4.2.7 of [**Gra09**]). Moreover, the explicit computation allows to observe that this subgroup corresponds to the usual q-difference Galois group as described in [**Sau04a**] of the constant linear q-difference system $X(qx) = AX(x)$ (*cf.* Theorem 4.4.2 of [**Gra09**] or Theorem 2.4 of [**Gra12**]).

CHAPTER 9

Comparison of the parametrized intrinsic Galois group with the Galois D-groupoid

A. Granier has defined a D-groupoid for non-linear q-difference equations, in analogy with Malgrange D-groupoid for non-linear differential equations (see the previous chapter). Roughly, this D-groupoid corresponds to the largest sheaf of analytic differential equations that kill the dynamics of the non-linear q-difference equation.

In this section we prove that the Malgrange-Granier D-groupoid, in the special case of a linear q-difference equation, essentially "coincides" with the parametrized intrinsic Galois group of the equation. This result, which is Corollary 9.12, is not a priori straightforward because one has to compare a D-groupoid defined as a sheaf of differential ideal over an analytic variety and a ∂-group scheme *à la Kolchin*. This answers a question of Malgrange ([**Mal09**], page 2]).

Our proof is divided in three main steps. The first one relies on Theorem 7.13 and allows us to compare the parametrized intrinsic Galois group with the smallest ∂-scheme that contains the dynamic, namely its Kolchin closure. Then, we sheafify the defining equations of the Kolchin closure in order to get an algebraic D-groupoid, which is defined by the largest set of algebraic differential equations that kill the dynamic. Finally thanks to GAGA arguments, we show that the defining equations of the Malgrange-Granier D-groupoid are global and algebraic and thus coincide with the ones of our algebraic D-groupoid. In the differential case, the problem of the algebraicity of the D-groupoid has been tackled in more recent works by B. Malgrange himself.

In the special case of a linear differential equation, Malgrange proves that his D-groupoid, allows to recover the Picard-Vessiot group (see [**Mal01**]). The foliation associated to the solutions of the non-linear differential equation, which exists due to the Cauchy theorem, plays a central role in his proof, and actually in the whole theory. There is a true hindrance to prove a Cauchy theorem and define a foliation over $\mathbb{C}$ attached to a q-difference system. First of all, the solutions of a q-difference equation must be defined over a q-invariant domain and they usually have an essential singularity at 0 and at ∞. This fact prevents the existence of a local solution on a compact domain and therefore a transposition of the Cauchy theorem. To overcome the lack of local solutions, we use Theorem 7.13 as a crucial ingredient of our proof. However, some steps of our proof are similar to Malgrange theorem (*cf.* [**Mal01**]) and Granier's proof in the case of q-difference system with constant coefficients (see [**Gra12**], §2.1]). In §9.4 below, we show how in Malgrange or Granier's former comparison results, a parametrized intrinsic Galois group is hidden and why the parametrized structure is inherent to Malgrange's D-groupoid constructions.

Our results shall give some hints to compare the algebraic definitions of Morikawa of the Galois group of a non-linear q-difference equation and the analytic definitions of A.Granier (*cf.* [**Mor09**], [**MU09**], [**Ume10**]).

9.1. The Kolchin closure of the Dynamics and the Malgrange-Granier groupoid

Let $q \in \mathbb{C}^*$ be not a root of unity and let $A(x) \in \mathrm{GL}_\nu(\mathbb{C}(x))$. We consider the linear q-difference system

$$(9.1) \qquad Y(qx) = A(x)Y(x).$$

We set:

$$A_k(x) := A(q^{k-1}x)\ldots A(qx)A(x) \text{ for all } k \in \mathbb{Z}, k > 0;$$
$$A_0(x) = Id_\nu$$
$$A_k(x) := A(q^k x)^{-1} A(q^{k+1}x)^{-1}\ldots A(q^{-1}x)^{-1} \text{ for all } k \in \mathbb{Z}, k < 0,$$

so that $Y(q^k x) = A_k(x)Y(x)$, for any $k \in \mathbb{Z}$. Following Chapter 8, we denote by M the analytic complex variety $\mathbb{P}^1_\mathbb{C} \times \mathbb{C}^\nu$, by $\mathcal{G}al(A(x))$ the Galois D-groupoid of the system (9.1), i.e., the D-envelop of the dynamics

$$(9.2) \qquad Dyn(A(x)) = \left\{(x,X) \longmapsto (q^k x, A_k(x)X) : k \in \mathbb{Z}\right\}$$

in the space of jets $J^*(M,M)$. We keep the notation of Chapter 8, which is preliminary to the content of this section.

WARNING 9.1. Following Malgrange and the convention in Chapter 8, we say that a D-groupoid $\mathcal{H}$ is contained in a D-groupoid $\mathcal{G}$ if the groupoid of solutions of $\mathcal{H}$ is contained in the groupoid of solutions of $\mathcal{G}$. We will write $sol(\mathcal{H}) \subset sol(\mathcal{G})$ or equivalently $\mathcal{I}_\mathcal{G} \subset \mathcal{I}_\mathcal{H}$, where $\mathcal{I}_\mathcal{G}$ and $\mathcal{I}_\mathcal{H}$ are the (sheaves of) ideals of definition of $\mathcal{G}$ and $\mathcal{H}$, respectively.

NOTATION 9.2. In this section we introduce many tools that we use to get the proof of our main result Corollary 9.12. For the reader convenience we make a list of them here, with the reference to their definitions:

$Dyn(A(x))$,	(9.2);		
$\mathcal{G}al(A(x))$,	§8.2;	$\widetilde{\mathcal{G}al(A(x))}$,	Definition 8.8;
$\mathcal{G}al^{alg}(A(x))$,	Definition 9.3;	$\widetilde{\mathcal{G}al^{alg}(A(x))}$,	Definition 9.8;
$\mathcal{K}ol(A(x))$,	Definition 9.3;	$\widetilde{\mathcal{K}ol(A(x))}$,	Definition 9.5;
$\mathcal{L}in$,	Proposition 8.2;	$\mathcal{T}rv$,	Definition 8.7.

9.2. The groupoid $\mathcal{G}al^{alg}(A(x))$

Let $\mathbb{C}(x)\left\{T, \frac{1}{\det T}\right\}_\partial$, with $T = (T_{i,j} : i,j = 0,1,\ldots,\nu)$, be the algebra of differential rational functions over $\mathrm{GL}_{\nu+1}(\mathbb{C}(x))$. We consider the following morphism of ∂-$\mathbb{C}[x]$-algebras

$$\tau: \quad \mathbb{C}[x]\left\{T, \tfrac{1}{\det T}\right\}_\partial \quad \longrightarrow \quad H^0(M \times_\mathbb{C} M, \mathcal{O}_{J^*(M,M)})$$

$$\begin{pmatrix} T_{0,0} & T_{0,1} & \cdots & T_{0,\nu} \\ T_{1,0} & & & \\ \vdots & & (T_{i,j})_{i,j} & \\ T_{\nu,0} & & & \end{pmatrix} \longmapsto \begin{pmatrix} \frac{\partial \overline{x}}{\partial x} & \frac{\partial \overline{x}}{\partial X_1} & \cdots & \frac{\partial \overline{x}}{\partial X_\nu} \\ \frac{\partial \overline{X_1}}{\partial x} & & & \\ \vdots & & \left(\frac{\partial \overline{X_i}}{\partial X_j}\right)_{i,j} & \\ \frac{\partial \overline{X_\nu}}{\partial x} & & & \end{pmatrix}$$

from $\mathbb{C}[x]\left\{T, \tfrac{1}{\det T}\right\}_\partial$ to the global sections $H^0(M \times_\mathbb{C} M, \mathcal{O}_{J^*(M,M)})$ of $\mathcal{O}_{J^*(M,M)}$, that can be thought as the algebra of global partial differential equations over $M \times M$. The image by τ of the differential ideal

$$\mathcal{I} = (T_{0,1}, \ldots, T_{0,\nu}, T_{1,0}, \ldots, T_{\nu,0}, \partial(T_{0,0})),$$

that defines the ∂-group scheme

$$\left\{ diag(\alpha, \beta(x)) := \begin{pmatrix} \alpha & 0 \\ 0 & \beta(x) \end{pmatrix} \,:\, \text{where } \alpha \in \mathbb{C}^* \text{ and } \beta(x) \in \operatorname{GL}_\nu(\mathbb{C}(x)) \right\},$$

is contained in the ideal $\mathcal{I}_{\mathcal{L}in}$ defining the D-groupoid $\mathcal{L}in$ (cf. Proposition 8.2).

DEFINITION 9.3. We call $\mathcal{K}ol(A(x))$ the smallest ∂-$\mathbb{C}(x)$-scheme of $\operatorname{GL}_{\nu+1}(\mathbb{C}(x))$, defined over $\mathbb{C}(x)$, which contains

$$\left\{ diag(q^k, A_k(x)) := \begin{pmatrix} q^k & 0 \\ 0 & A_k(x) \end{pmatrix} \,:\, k \in \mathbb{Z} \right\},$$

and has the following property: if we call $I_{\mathcal{K}ol(A(x))}$ the differential ideal defining $\mathcal{K}ol(A(x))$ and $I'_{\mathcal{K}ol(A(x))} = I_{\mathcal{K}ol(A(x))} \cap \mathbb{C}[x]\left\{T, \tfrac{1}{\det T}\right\}_\partial$, then the (sheaf of) differential ideal $\langle \mathcal{I}_{\mathcal{L}in}, \tau(I'_{\mathcal{K}ol(A(x))}) \rangle$ generates a D-groupoid, that we will call $\mathcal{G}al^{alg}(A(x))$, in the space of jets $J^*(M, M)$.

REMARK 9.4. The definition above requires some explanations:
- For the reader convenience, we recall here the basic definitions of the theory of affine differential schemes, that can be found in [**Kov02**]. If we fix a ∂-field k of characteristic zero, we define a ∂-k-scheme as follows: An affine ∂-k-scheme (or ∂-scheme over k) is a (covariant) functor from the category of ∂-k-algebras to the category of sets which is representable. It means that a functor X from the category of ∂-k-algebras to the category of sets is a ∂-k-scheme if and only if there exists a ∂-k-algebra $k\{X\}$ and an isomorphism of functors $X \simeq \operatorname{Alg}_k^\partial(k\{X\}, -)$, where $\operatorname{Alg}_k^\partial$ stands for morphism of ∂-k-algebras. By the Yoneda lemma, the ∂-k-algebra $k\{X\}$ is uniquely determined up to unique ∂-k-isomorphisms. We call it the ring of ∂-coordinates of X. By a closed ∂-k-subscheme $Y \subset X$ we mean a subfunctor Y of X which is represented by $k\{X\}/\mathbb{I}(Y)$ for some ∂-ideal $\mathbb{I}(Y)$ of $k\{X\}$. The ideal $\mathbb{I}(Y)$ of $k\{X\}$ is uniquely determined by Y and vice versa. We call it the vanishing ideal of Y in X. A morphism of ∂-k-schemes is a morphism of functors. If $\phi \colon X \to Y$ is a morphism of ∂-k-schemes, we denote the dual morphism of ∂-k-algebras with $\phi^* \colon k\{Y\} \to k\{X\}$.

  Reduced ∂-schemes correspond to differential varieties in the sense of Kolchin (see for instance [**Kol73**]), for whom it suffices to focus on the

solution set of a system of differential equations with value in a sufficiently big field, i.e., a ∂-closed field.
- The phrase "smallest ∂-$\mathbb{C}(x)$-subscheme of $\mathrm{GL}_{\nu+1}(\mathbb{C}(x))$" must be understood in the following way. The ideal of definition of $\mathcal{K}ol(A(x))$ is the largest differential ideal of $\mathbb{C}(x)\left\{T,\frac{1}{\det T}\right\}_\partial$ which admits the matrices $diag(q^k, A_k(x))$ as solutions for any $k \in \mathbb{Z}$ and verifies the second requirement of the definition. Then $I_{\mathcal{K}ol(A(x))}$ is radical and the Ritt-Raudenbush theorem (cf. Theorem 7.7 above) implies that $I_{\mathcal{K}ol(A(x))}$ is finitely ∂-generated. Of course, the $\mathbb{C}(x)$-rational points of $\mathcal{K}ol(A(x))$ may give very poor information on its structure, so we would rather speak of solutions in a differential closure of $\mathbb{C}(x)$.
- The structure of D-groupoid has the following consequence on the points of $\mathcal{K}ol(A(x))$: if $diag(\alpha, \beta(x))$ and $diag(\gamma, \delta(x))$ are two matrices with entries in a differential extension of $\mathbb{C}(x)$ that belong to $\mathcal{K}ol(A(x))$ then the matrix $diag(\alpha\gamma, \beta(\gamma x)\delta(x))$ belongs to $\mathcal{K}ol(A(x))$. In other words, the set of local diffeomorphisms $(x, X) \mapsto (\alpha x, \beta(x)X)$ of $M \times M$ such that $diag(\alpha, \beta(x))$ belongs to $\mathcal{K}ol(A(x))$ forms a set theoretic groupoid. We could have supposed only that $\mathcal{K}ol(A(x))$ is a ∂-$\mathbb{C}(x)$-scheme and the solutions of $\mathcal{K}ol(A(x))$ form a groupoid in the sense above, but this wouldn't have been enough. In fact, it is not known if a sheaf of differential ideals of $J^*(M, M)$ whose solutions forms a groupoid is actually a D-groupoid (cf. Definition 8.1, and in particular conditions (ii') and (iii')). B. Malgrange told us that he can only prove this statement for a Lie algebra.

The ∂-$\mathbb{C}(x)$-scheme $\mathcal{K}ol(A(x))$ is going to be a bridge between the parametrized intrinsic Galois group and the Galois D-groupoid $\mathcal{G}al(A(x))$ defined in the previous chapter, via the following theorem.

DEFINITION 9.5. Let $\mathcal{M}_{\mathbb{C}(x)}^{(A)} := (\mathbb{C}(x)^\nu, \Sigma_q : X \mapsto A^{-1}\sigma_q(X))$ be the q-difference module over $\mathbb{C}(x)$ associated to the system $Y(qx) = A(x)Y(x)$, where $\sigma_q(X)$ is defined componentwise. We call $\widetilde{\mathcal{K}ol(A(x))}$ the ∂-$\mathbb{C}(x)$-group scheme defined by the differential ideal $\langle I_{\mathcal{K}ol(A(x))}, T_{0,0} - 1\rangle$ in $\mathbb{C}(x)\left\{T, \frac{1}{\det T}\right\}_\partial$.

Notice that, as for the Zariski closure, the Kolchin closure does not commute with the intersection, therefore $\widetilde{\mathcal{K}ol(A(x))}$ is not the Kolchin closure of $\{A_k(x)\}_k$. Then we have:

THEOREM 9.6. $Gal^\partial(\mathcal{M}_{\mathbb{C}(x)}^{(A)}) \cong \widetilde{\mathcal{K}ol(A(x))}$.

REMARK 9.7. One can define in exactly the same way a $\mathbb{C}(x)$-subscheme $\mathcal{Z}ar(A)$ of $\mathrm{GL}_{\nu+1}(\mathbb{C}(x))$ containing the dynamics of the system and such that
$$\{(x, X) \mapsto (\alpha x, \beta(x)X) : diag(\alpha, \beta(x)) \in \mathcal{Z}ar(A)\}$$
is a subgroupoid of the groupoid of diffeomorphisms of $M \times M$. Then one proves in the same way that $\widetilde{\mathcal{Z}ar(A)}$ coincide with the intrinsic Galois group, introduced in Chapter 6.

PROOF OF THEOREM 9.6. Let $\mathcal{N} = constr^\partial(\mathcal{M})$ be a construction of differential algebra of $\mathcal{M}$. We can consider:

- The basis denoted by $constr^\partial(\underline{e})$ of $\mathcal{N}$ built from the canonical basis $\underline{e}$ of $\mathbb{C}(x)^\nu$, applying the same constructions of linear differential algebra.
- For any $\beta \in \mathrm{GL}_\nu(\mathbb{C}(x))$, the matrix $constr^\partial(\beta)$ acting on $\mathcal{N}$ with respect to the basis $constr^\partial(\underline{e})$, obtained from β by functoriality. Its coefficients lies in $\mathbb{C}(x)[\beta, \partial(\beta), \dots]$
- Any $\psi = (\alpha, \beta) \in \mathbb{C}^* \times \mathrm{GL}_\nu(\mathbb{C}(x))$ acts semilinearly on $\mathcal{N}$ in the following way: $\psi \underline{e} = (constr^\partial(\beta))^{-1}\underline{e}$ and $\phi(f(x)n) = f(\alpha x)n$, for any $f(x) \in \mathbb{C}(x)$ and $n \in \mathcal{N}$. In particular, $(q^k, A_k(x)) \in \mathbb{C}^* \times \mathrm{GL}_\nu(\mathbb{C}(x))$ acts as Σ_q^k on $\mathcal{N}$.

A q-difference submodule $\mathcal{E}$ of $\mathcal{N}$ correspond to an invertible matrix $F \in \mathrm{GL}_\nu(\mathbb{C}(x))$ such that

$$(9.3) \qquad F(q^k x)^{-1} constr^\partial(A_k) F(x) = \begin{pmatrix} * & * \\ 0 & * \end{pmatrix}, \text{ for any } k \in \mathbb{Z}.$$

Now, $(1, \beta) \in \mathbb{C}^* \times \mathrm{GL}_\nu(\mathbb{C}(x))$ stabilizes $\mathcal{E}$ if and only if

$$(9.4) \qquad F(x)^{-1} constr^\partial(\beta) F(x) = \begin{pmatrix} * & * \\ 0 & * \end{pmatrix}.$$

Equation (9.3) corresponds to a differential polynomial $L(T_{0,0}, (T_{i,j})_{i,j \geq 1})$ belonging to $\mathbb{C}(x)\{T, \frac{1}{\det T}\}_\partial$ and having the property that $L(q^k, (A_k)) = 0$, for all $k \in \mathbb{Z}$. On the other hand (9.4) corresponds to $L(1, (T_{i,j})_{i,j \geq 1}))$. It means that the solutions of the differential ideal $\langle I_{\mathcal{K}ol(A(x))}, T_{0,0} - 1 \rangle \subset \mathbb{C}(x)\{T, \frac{1}{\det T}\}_\partial$ stabilize all the q-difference submodules of all the constructions of differential algebra, and hence that

$$\widetilde{\mathcal{K}ol(A(x))} \subset Gal^\partial(\mathcal{M}_{\mathbb{C}(x)}).$$

Let us prove the inverse inclusion. In the notation of Theorem 7.19, there exists a finitely generated extension K of $\mathbb{Q}$ and a σ_q-stable subalgebra $\mathcal{A}$ of $K(x)$ of the forms considered in §7.3 such that:

(1) $A(x) \in \mathrm{GL}_\nu(\mathcal{A})$, so that it defines a q-difference module $\mathcal{M}_{K(x)}^{(A)}$ over $K(x)$;
(2) $Gal^\partial(\mathcal{M}_{K(x)}^{(A)}) \otimes_{K(x)} \mathbb{C}(x) \cong Gal^\partial(\mathcal{M}_{\mathbb{C}(x)}^{(A)})$;
(3) $\mathcal{K}ol(A(x))$ is defined over $\mathcal{A}$, i.e., there exists a differential ideal I in the differential ring $\mathcal{A}\{T, \frac{1}{\det(T)}\}_\partial$ such that I generates $I_{\mathcal{K}ol(A(x))}$ in $\mathbb{C}(x)\{T, \frac{1}{\det T}\}_\partial$.

For any element $\widetilde{L}$ of the defining ideal of $\widetilde{\mathcal{K}ol(A(x))}$ over $\mathcal{A}$, there exists

$$L(T_{0,0}; T_{i,j}, i, j = 1, \dots, \nu) \in I \subset \mathcal{A}\left\{T, \frac{1}{\det(T)}\right\}_\partial,$$

such that $L \in \mathcal{I}_{\mathcal{K}ol(A(x))}$ and $\widetilde{L} = L(1; T_{i,j}, i, j = 1, \dots, \nu)$. If q is an algebraic number, other than a root of unity, or if q is transcendental, then, for almost all places $v \in \mathcal{C}$, we have

$$\widetilde{L}(A_{\kappa_v}) \equiv L(1, A_{\kappa_v}) \equiv L(q^{\kappa_v}, A_{\kappa_v}) \equiv 0 \text{ modulo } \phi_v.$$

This shows that $\widetilde{\mathcal{K}ol(A(x))}$ is a ∂-$\mathbb{C}(x)$-subgroup scheme of $\mathrm{GL}_\nu(\mathbb{C}(x))$ which contains a non-empty cofinite set of v-curvatures, in the sense of Theorem 7.19. Therefore, $\widetilde{\mathcal{K}ol(A(x))}$ contains the parametrized intrinsic Galois group of $\mathcal{M}_{\mathbb{C}(x)}^{(A)}$. $\square$

DEFINITION 9.8. We call $\widetilde{\mathcal{G}al^{alg}}(A(x))$ the intersection of $\mathcal{G}al^{alg}(A(x))$ and $\mathcal{T}rv$.

It follows from the definition that the D-groupoid $\widetilde{\mathcal{G}al^{alg}}(A(x))$ is generated by its global equations, i.e., by $\mathcal{L}in$ and the image of the equations of $\widetilde{\mathcal{K}ol}(A(x))$ by the morphism τ. Therefore we deduce from Theorem 9.6 the following statement:

COROLLARY 9.9. *As a D-groupoid, $\widetilde{\mathcal{G}al^{alg}}(A(x))$ is generated by its global sections, namely the D-groupoid $\mathcal{L}in$ and the image of the equations of $Gal^\partial(\mathcal{M}_{\mathbb{C}(x)}^{(A)})$ via the morphism τ.*

REMARK 9.10. The corollary above says that the sheaf of differential ideals defining the D-groupoid $\widetilde{\mathcal{G}al^{alg}}(A(x))$ is generated by its global sections, $\mathcal{L}in$ and $\mathfrak{q}$, where $\mathfrak{q}$ is the defining ideal of the intrinsic parametrized Galois group. This statement is much stronger than saying that, in the neighborhood of $x_0 \in \mathbf{P}^1(\mathbb{C})$, the germs of diffeomorphism, solutions of $\widetilde{\mathcal{G}al^{alg}}(A(x))$, can be written $(x, X) \mapsto (x, \beta(x)X)$ with $\beta(x) \in \mathrm{GL}_\nu(\mathbb{C}\{x - x_0\})$ solution of $\mathfrak{q}$.

The D-groupoid $\widetilde{\mathcal{G}al^{alg}}(A(x))$ is a differential analog of the D-groupoid generated by a group scheme introduced in [**Mal01**, Proposition 5.3.2] by B. Malgrange.

9.3. The Galois D-groupoid $\mathcal{G}al(A(x))$ vs the intrinsic parametrized Galois group

Since $Dyn(A(x))$ is contained in the solutions of $\mathcal{G}al^{alg}(A(x))$, we have

$$sol(\mathcal{G}al(A(x))) \subset sol(\mathcal{G}al^{alg}(A(x)))$$

and

$$sol(\widetilde{\mathcal{G}al(A(x))}) \subset sol(\widetilde{\mathcal{G}al^{alg}}(A(x))).$$

As already mentioned, the solution are to be found in some differential closure of $(\mathbb{C}(x), \partial)$.

THEOREM 9.11. *The solutions of the D-groupoid $\widetilde{\mathcal{G}al}(A(x))$ (resp. $\mathcal{G}al(A(x))$) coincide with the solutions of $\widetilde{\mathcal{G}al^{alg}}(A(x))$ (resp. $\mathcal{G}al^{alg}(A(x))$).*

Combining the theorem above with Corollary 9.9, we immediately obtain:

COROLLARY 9.12. *The solutions of the D-groupoid $\widetilde{\mathcal{G}al}(A(x))$ are germs of diffeomorphisms of the form $(x, X) \mapsto (x, \beta(x)X)$, such that $\beta(x)$ is a solution of the differential equations defining $Gal^\partial(\mathcal{M}_{\mathbb{C}(x)}^{(A)})$, and vice versa.*

REMARK 9.13. The corollary above says that the germs of diffeomorphism, solutions of $\widetilde{\mathcal{G}al}(A(x))$, in a neighborhood of a transversal $\{x_0\} \times \mathbb{C}^\nu$ (*cf.* Proposition 8.10 above) are of the form $(x, X) \mapsto (x, \beta(x)X)$ with $\beta(x) \in \mathrm{GL}_\nu(\mathbb{C}\{x - x_0\})$ of the differential equations defining the parametrized intrinsic Galois group. Notice that we do not say that the sheaf of differential ideals of $\widetilde{\mathcal{G}al}(A(x))$ is generated by $\mathcal{L}in$ and the defining equations of the intrinsic parametrized Galois group, which would be a stronger statement.

9.3. THE GALOIS D-GROUPOID VS THE INTRINSIC PARAMETRIZED GALOIS GROUP

PROOF OF THEOREM 9.11. Let $\mathcal{I}$ be the differential ideal of $\mathcal{G}al(A(x))$ in $\mathcal{O}_{J^*(M,M)}$ and let $\mathcal{I}_r$ be the subideal of $\mathcal{I}$ of order $\leq r$. We consider the morphism of analytic varieties given by

$$\iota: \mathbb{P}^1_{\mathbb{C}} \times \mathbb{P}^1_{\mathbb{C}} \longrightarrow M \times_{\mathbb{C}} M$$
$$(x, \overline{x}) \longmapsto (x, 0, \overline{x}, 0)$$

and the inverse image $\mathcal{J}_r := \iota^{-1}\mathcal{I}_r$ (resp. $\mathcal{J} := \iota^{-1}\mathcal{I}$) of the sheaf $\mathcal{I}_r$ (resp. $\mathcal{I}$) over $\mathbb{P}^1_{\mathbb{C}} \times \mathbb{P}^1_{\mathbb{C}}$. We consider similarly to [**Mal01**, Lemma 5.3.3], the evaluation $ev(\iota^{-1}\mathcal{I})$ at $X = \overline{X} = \frac{\partial^i \overline{X}}{\partial x^i} = 0$ of the equations of $\iota^{-1}\mathcal{I}$ and we denote by $ev(\mathcal{I})$ the direct image by ι of the sheaf $ev(\iota^{-1}\mathcal{I})$.

The following lemma is crucial in the proof of the Theorem 9.11:

LEMMA 9.14. *A germ of local diffeomorphism $(x, X) \mapsto (\alpha x, \beta(x) X)$ of M is solution of $\mathcal{I}$ if and only if it is solution of $ev(\mathcal{I})$.*

PROOF. First of all, we notice that $\mathcal{I}$ is contained in $\mathcal{L}in$. Moreover the solutions of $\mathcal{I}$, that are diffeomorphisms mapping a neighborhood of $(x_0, X_0) \in M$ to a neighborhood of $(\overline{x}_0, \overline{X}_0)$, can be naturally continued to diffeomorphisms of a neighborhood of $x_0 \times \mathbb{C}^\nu$ to a neighborhood of $\overline{x}_0 \times \mathbb{C}^\nu$. Therefore it follows from the particular structure of the solutions of $\mathcal{L}in$, that they are also solutions of $ev(\mathcal{I})$ (*cf.* Proposition 8.2).

Conversely, let the germ of diffeomorphism $(x, X) \mapsto (\alpha x, \beta(x) X)$ be a solution of $ev(\mathcal{I})$ and $E \in \mathcal{I}_r$. It follows from Proposition 8.4 that there exists $E_1 \in \mathcal{I}$ of order r, only depending on the variables $x, X, \frac{\partial \overline{x}}{\partial x}, \frac{\partial \overline{X}}{\partial X}, \frac{\partial^2 \overline{X}}{\partial x \partial X}, \ldots, \frac{\partial^r \overline{X}}{\partial x^{r-1} \partial X}$, such that $(x, X) \mapsto (\alpha x, \beta(x) X)$ is solution of E if and only if it is solution of E_1. So we will focus on equations on the form E_1 and, to simplify notation, we will write E for E_1.

By assumption $(x, X) \mapsto (\alpha x, \beta(x) X)$ is solution of

$$E\left(x, 0, \frac{\partial \overline{x}}{\partial x}, \frac{\partial \overline{X}}{\partial X}, \frac{\partial^2 \overline{X}}{\partial x \partial X}, \ldots \frac{\partial^r \overline{X}}{\partial x^{r-1} \partial X}\right)$$

and we have to show that $(x, X) \mapsto (\alpha x, \beta(x) X)$ is a solution of E. We consider the Taylor expansion of E:

$$E\left(x, X, \frac{\partial \overline{x}}{\partial x}, \frac{\partial \overline{X}}{\partial X}, \frac{\partial^2 \overline{X}}{\partial x \partial X}, \ldots \frac{\partial^r \overline{X}}{\partial x^{r-1} \partial X}\right) = \sum_\alpha E_\alpha(x, X) \partial^\alpha,$$

where ∂^α is a monomial in the coordinates $\frac{\partial \overline{x}}{\partial x}, \frac{\partial \overline{X}}{\partial X}, \frac{\partial^2 \overline{X}}{\partial x \partial X}, \ldots \frac{\partial^r \overline{X}}{\partial x^{r-1} \partial X}$. Developing the $E_\alpha(x, X)$ with respect to $X = (X_1, \ldots, X_\nu)$ we obtain:

$$E = \sum_{\underline{k}} \left(\sum_\alpha \left(\frac{\partial^{\underline{k}} E_\alpha}{\partial X^{\underline{k}}}\right)(x, 0) \partial^\alpha \right) X^{\underline{k}},$$

with $\underline{k} \in (\mathbb{Z}_{\geq 0})^\nu$. If we show that for any $\underline{k}$ the germ $(x, X) \mapsto (\alpha x, \beta(x) X)$ verifies the equation

$$B_{\underline{k}} := \sum_\alpha \left(\frac{\partial^{\underline{k}} E_\alpha}{\partial X^{\underline{k}}}\right)(x, 0) \partial^\alpha$$

we can conclude. For $\underline{k} = (0, \ldots, 0)$, there is nothing to prove since $B_{\underline{0}} = ev(E)$.

Let D_{X_i} be the derivation of $\mathcal{I}$ corresponding to $\frac{\partial}{\partial X_i}$, The differential equation

$$D_{X_i}(E) = \sum_\alpha \left(\frac{\partial E_\alpha}{\partial X_i}\right)(x,X)\partial^\alpha + \sum_\alpha E_\alpha(x,X) D_{X_i}(\partial^\alpha)$$

is still in $\mathcal{I}$, since $\mathcal{I}$ is a differential ideal. Therefore by assumption $(x,X) \mapsto (\alpha x, \beta(x)X)$ is a solution of

$$ev(D_{X_i} E) = \sum_\alpha \left(\frac{\partial E_\alpha}{\partial X_i}\right)(x,0)\partial^\alpha + \sum_\alpha E_\alpha(x,0) D_{X_i}(\partial^\alpha).$$

Since $D_{X_i}(\partial^\alpha) \in \mathcal{L}in$ and $(x,X) \mapsto (\alpha x, \beta(x)X)$ is a solution of $\mathcal{L}in$, we conclude that $(x,X) \mapsto (\alpha x, \beta(x)X)$ is a solution of

$$\sum_\alpha \left(\frac{\partial E_\alpha}{\partial X}\right)(x,0)\partial^\alpha$$

and therefore of B_1. Iterating the argument, one deduce that $(x,X) \mapsto (\alpha x, \beta(x)X)$ is solution of $B_{\underline{k}}$ for any $\underline{k} \in (\mathbb{Z}_{\geq 0})^\nu$, which ends the proof of the lemma. $\square$

We go back to the proof of Theorem 9.11. Lemma 9.14 proves that the solutions of $\mathcal{G}al(A(x))$ coincide with those of the D-groupoid Γ generated by $\mathcal{L}in$ and $ev(\mathcal{I})$, defined on the open neighborhoods of any $x_0 \times \mathbb{C}^\nu \in M$. By intersection with the equation $\mathcal{T}rv$, the same holds for the transversal groupoids $\widetilde{\mathcal{G}al(A(x))}$ and $\widetilde{\Gamma}$.

Since $\mathbb{P}^1_\mathbb{C} \times \mathbb{P}^1_\mathbb{C}$ and $M \times_\mathbb{C} M$ are locally compact and $\mathcal{I}_r$ is a coherent sheaf over $M \times_\mathbb{C} M$, the sheaf $\mathcal{J}_r$ is an analytic coherent sheaf over $\mathbb{P}^1_\mathbb{C} \times \mathbb{P}^1_\mathbb{C}$ and so is its quotient $ev(\iota^{-1}(\mathcal{I}_r))$. By [**Ser56**, Theorem 3], there exists an algebraic coherent sheaf $\mathbb{J}_r$ over the projective variety $\mathbb{P}^1_\mathbb{C} \times \mathbb{P}^1_\mathbb{C}$ such that the analyzation of $\mathbb{J}_r$ coincides with $ev(\iota^{-1}(\mathcal{I}_r))$. This implies that $ev(\mathcal{I})$ is generated by algebraic differential equations which by definition have the dynamics for solutions.

We thus have that the $sol(\Gamma) = sol(\mathcal{G}al(A(x))) \subset sol(\mathcal{G}al^{alg}(A(x)))$. Since both Γ and $\mathcal{G}al^{alg}(A(x))$ are algebraic, the minimality of the variety $\mathcal{K}ol(A(x))$ implies that $sol(\mathcal{G}al^{alg}(A(x))) \subset sol(\Gamma)$. We conclude that the solutions of $\mathcal{G}al(A(x))$ coincide with those $\mathcal{G}al^{alg}(A(x))$. The same hold for $\widetilde{\mathcal{G}al(A(x))}$, $\widetilde{\Gamma}$ and $\widetilde{\mathcal{G}al^{alg}(A(x))}$. This concludes the proof. $\square$

9.4. Comparison with known results

In [**Mal01**], B. Malgrange proves that the Galois-D-groupoid of a linear differential equation allows to recover, in the special case of a linear differential equation, the Picard-Vessiot group over $\mathbb{C}$. This is compatible with the result above, since:

- due to the fact that local solutions of a linear differential equation form a $\mathbb{C}$-vector space (rather than a vector space on the field of elliptic functions!), [**Kat82**, Proposition 4.1] shows that the intrinsic Galois group and the Picard-Vessiot group in the differential setting become isomorphic above a certain extension of the local ring. For more details on the relation between the intrinsic Galois group and the usual Galois group see [**Pil02**, Corollary 3.3].
- In the differential setting the parametrized intrinsic Galois group with respect to $\frac{d}{dx}$ is conjugate to the constant points of the differential group scheme attached to the intrinsic Galois group. Therefore, the set of germs

9.4. COMPARISON WITH KNOWN RESULTS

of solutions of the defining equations of the parametrized intrinsic Galois group coincide with the $\mathbb{C}$-points of the intrinsic Galois group.

Therefore B. Malgrange actually finds a parametrized intrinsic Galois group, which is hidden in his construction. The steps of the proof above are the same as in his proof, apart that, to compensate the lack of good local solutions, we are obliged to use Theorem 7.13. Anyway, the application of Theorem 7.13 appears to be very natural, if one considers how close the definition of the dynamics of a linear q-difference system and the definition of the curvatures are.

In [**Gra12**], A. Granier shows that in the case of a q-difference system with constant coefficients the groupoid that fixes the transversals in $\mathcal{G}al(A(x))$ is the Picard-Vessiot group, i.e., a $\mathbb{C}$-group scheme. Once again, this is not in contradiction with our results. Indeed, let $\mathcal{M}$ be a q-difference module over $\mathbb{C}(x)$ associated with a constant q-difference system. Under this assumption, the curvatures of the system are defined over $\mathbb{C}$. Thus, the intrinsic Galois group is defined over $\mathbb{C}$ and coincides with the Picard-Vessiot group. Moreover since the prolongation of $\mathcal{M}$ splits, the parametrized intrinsic Galois group is conjugate to the group of constant points of the differential group scheme attached to the intrinsic Galois group. This allows to conclude that the set of germs of solutions of the defining equations of the parametrized intrinsic Galois group coincides with the Picard-Vessiot group for a constant q-difference system.

Because of these results, G. Casale and J. Roques have conjectured that "for linear (q-)difference systems, the action of Malgrange groupoid on the fibers gives the classical Galois groups" (*cf.* [**CR08**]). In *loc. cit.*, they give two proofs of their main integrability result: one of them relies on their conjecture. Here we have proved that the Galois-D-groupoid allows to recover exactly the parametrized intrinsic Galois group. By taking the Zariski closure one can also recover the intrinsic Galois group. One can prove that we can also recover the Picard-Vessiot group (*cf.* [**vdPS97**], [**Sau04a**]), performing a Zariski closure and a suitable field extension, and the parametrized Galois group (*cf.* [**HS08**]), performing a field extension. See Remarks 6.17 and 7.20.

Bibliography

[And01]   Y. André, *Différentielles non commutatives et théorie de Galois différentielle ou aux différences* (French, with English and French summaries), Ann. Sci. École Norm. Sup. (4) **34** (2001), no. 5, 685–739, DOI 10.1016/S0012-9593(01)01074-6. MR1862024

[And04]   Y. André, *Sur la conjecture des p-courbures de Grothendieck-Katz et un problème de Dwork* (French, with French summary), Geometric aspects of Dwork theory. Vol. I, II, Walter de Gruyter, Berlin, 2004, pp. 55–112. MR2023288

[AR13]   S. A. Abramov and A. A. Ryabenko, *Linear q-difference equations depending on a parameter*, J. Symbolic Comput. **49** (2013), 65–77, DOI 10.1016/j.jsc.2011.12.017. MR2997840

[Aut01]   P. Autissier, *Points entiers sur les surfaces arithmétiques* (French), J. Reine Angew. Math. **531** (2001), 201–235, DOI 10.1515/crll.2001.015. MR1810122

[BCDVW16] M. Barkatou, T. Cluzeau, J.-A. Weil, and L. Di Vizio, *Computing the Lie algebra of the differential Galois group of a linear differential system*, Proceedings of the 2016 ACM International Symposium on Symbolic and Algebraic Computation, ACM, New York, 2016, pp. 63–70, DOI 10.1145/2930889.2930932. MR3565698

[BCS14]   A. Bostan, X. Caruso, and É. Schost, *A fast algorithm for computing the characteristic polynomial of the p-curvature*, ISSAC 2014—Proceedings of the 39th International Symposium on Symbolic and Algebraic Computation, ACM, New York, 2014, pp. 59–66, DOI 10.1145/2608628.2608650. MR3239909

[BCS15]   A. Bostan, X. Caruso, and É. Schost, *A fast algorithm for computing the p-curvature*, ISSAC'15—Proceedings of the 2015 ACM International Symposium on Symbolic and Algebraic Computation, ACM, New York, 2015, pp. 69–76. MR3388284

[BCS16]   A. Bostan, X. Caruso, and É. Schost, *Computation of the similarity class of the p-curvature*, Proceedings of the 2016 ACM International Symposium on Symbolic and Algebraic Computation, ACM, New York, 2016, pp. 111–118, DOI 10.1145/2930889.2930897. MR3565704

[Bou64]   N. Bourbaki, *Éléments de mathématique. Fasc. XXX. Algèbre commutative. Chapitre 5: Entiers. Chapitre 6: Valuations* (French), Actualités Scientifiques et Industrielles [Current Scientific and Industrial Topics], No. 1308, Hermann, Paris, 1964. MR0194450

[BS09]   A. Bostan and É. Schost, *Fast algorithms for differential equations in positive characteristic*, ISSAC 2009—Proceedings of the 2009 International Symposium on Symbolic and Algebraic Computation, ACM, New York, 2009, pp. 47–54, DOI 10.1145/1576702.1576712. MR2742690

[Cas80]   F. Casorati, *Il calcolo delle differenze finite interpretato ed accresciuto di nuovi teoremi a sussidio principalmente delle odierne ricerche basate sulla variabilità complessa*, Annali di Matematica Pura ed Applicata, Series 2 (1880) **10** (1880), no. 1, 10–45.

[Cas72]   P. J. Cassidy, *Differential algebraic groups*, Amer. J. Math. **94** (1972), 891–954, DOI 10.2307/2373764. MR360611

[Coh65]   R. M. Cohn, *Difference algebra*, Interscience Publishers John Wiley & Sons, New York-London-Sydeny, 1965. MR0205987

[CR08]   G. Casale and J. Roques, *Dynamics of rational symplectic mappings and difference Galois theory*, Int. Math. Res. Not. IMRN, posted on 2008, Art. ID rnn 103, 23, DOI 10.1093/imrn/rnn103. MR2439539

[CS12a] S. Chen and M. F. Singer, *Residues and telescopers for bivariate rational functions*, Adv. in Appl. Math. **49** (2012), no. 2, 111–133, DOI 10.1016/j.aam.2012.04.003. MR2946428

[CS12b] S. Chen and M. F. Singer, *Residues and telescopers for bivariate rational functions*, Adv. in Appl. Math. **49** (2012), no. 2, 111–133, DOI 10.1016/j.aam.2012.04.003. MR2946428

[DG70] M. Demazure and P. Gabriel, *Groupes algébriques. Tome I: Géométrie algébrique, généralités, groupes commutatifs* (French), Masson & Cie, Éditeur, Paris; North-Holland Publishing Co., Amsterdam, 1970. Avec un appendice *Corps de classes local* par Michiel Hazewinkel. MR0302656

[DV02] L. Di Vizio, *Arithmetic theory of q-difference equations: the q-analogue of Grothendieck-Katz's conjecture on p-curvatures*, Invent. Math. **150** (2002), no. 3, 517–578, DOI 10.1007/s00222-002-0241-z. MR1946552

[DVH10] L. Di Vizio and C. Hardouin, *Courbures, groupes de Galois génériques et D-groupoïde de Galois d'un système aux q-différences* (French, with English and French summaries), C. R. Math. Acad. Sci. Paris **348** (2010), no. 17-18, 951–954, DOI 10.1016/j.crma.2010.08.001. MR2721777

[DVH12] L. Di Vizio and C. Hardouin, *Descent for differential Galois theory of difference equations: confluence and q-dependence*, Pacific J. Math. **256** (2012), no. 1, 79–104, DOI 10.2140/pjm.2012.256.79. MR2928542

[DVRSZ03] L. Di Vizio, J.-P. Ramis, J. Sauloy, and C. Zhang, *Équations aux q-différences* (French), Gaz. Math. **96** (2003), 20–49. MR1988639

[FRL06] C. Favre and J. Rivera-Letelier, *Équidistribution quantitative des points de petite hauteur sur la droite projective* (French, with English and French summaries), Math. Ann. **335** (2006), no. 2, 311–361, DOI 10.1007/s00208-006-0751-x. MR2221116

[GGO13] H. Gillet, S. Gorchinskiy, and A. Ovchinnikov, *Parameterized Picard-Vessiot extensions and Atiyah extensions*, Adv. Math. **238** (2013), 322–411, DOI 10.1016/j.aim.2013.02.006. MR3033637

[Gil02] H. Gillet, *Differential algebra—a scheme theory approach*, Differential algebra and related topics (Newark, NJ, 2000), World Sci. Publ., River Edge, NJ, 2002, pp. 95–123, DOI 10.1142/9789812778437_0003. MR1921696

[Gra09] A. Granier, *Un groupoïde de galois pour les équations aux q-différences*, Ph.D. thesis, Université Toulouse III Paul Sabatier, 2009.

[Gra10] A. Granier, *Un D-groupoïde de Galois local pour les systèmes aux q-différences fuchsiens* (French, with English and French summaries), C. R. Math. Acad. Sci. Paris **348** (2010), no. 5-6, 263–265, DOI 10.1016/j.crma.2010.01.030. MR2600119

[Gra12] A. Granier, *A Galois D-groupoid for q-difference equations* (English, with English and French summaries), Ann. Inst. Fourier (Grenoble) **61** (2011), no. 4, 1493–1516 (2012), DOI 10.5802/aif.2648. MR2951501

[Har10] C. Hardouin, *Iterative q-difference Galois theory*, J. Reine Angew. Math. **644** (2010), 101–144, DOI 10.1515/CRELLE.2010.053. MR2671776

[Hen96] P. Hendriks, *Algebraic Aspects of Linear Differential and Difference Equations*, Ph.D. thesis, University of Groningen., 1996.

[HS08] C. Hardouin and M. F. Singer, *Differential Galois theory of linear difference equations*, Math. Ann. **342** (2008), no. 2, 333–377, DOI 10.1007/s00208-008-0238-z. MR2425146

[HSS16] C. Hardouin, J. Sauloy, and M. F. Singer, *Galois theories of linear difference equations: an introduction*, Mathematical Surveys and Monographs, vol. 211, American Mathematical Society, Providence, RI, 2016. Papers from the courses held at the CIMPA Research School in Santa Marta, July 23–August 1, 2012, DOI 10.1090/surv/211. MR3410204

[Kam10] M. Kamensky, *Model theory and the Tannakian formalism*, Trans. Amer. Math. Soc. **367** (2015), no. 2, 1095–1120, DOI 10.1090/S0002-9947-2014-06062-5. MR3280038

[Kap57] I. Kaplansky, *An introduction to differential algebra*, Publ. Inst. Math. Univ. Nancago, No. 5, Hermann, Paris, 1957. MR0093654

[Kat70] N. M. Katz, *Nilpotent connections and the monodromy theorem: Applications of a result of Turrittin*, Inst. Hautes Études Sci. Publ. Math. **39** (1970), 175–232. MR291177

[Kat82]   N. M. Katz, *A conjecture in the arithmetic theory of differential equations* (English, with French summary), Bull. Soc. Math. France **110** (1982), no. 2, 203–239. MR667751

[Kol73]   E. R. Kolchin, *Differential algebra and algebraic groups*, Pure and Applied Mathematics, Vol. 54, Academic Press, New York-London, 1973. MR0568864

[Kov02]   J. J. Kovacic, *Differential schemes*, Differential algebra and related topics (Newark, NJ, 2000), World Sci. Publ., River Edge, NJ, 2002, pp. 71–94, DOI 10.1142/9789812778437_0002. MR1921695

[Lan83]   S. Lang, *Fundamentals of Diophantine geometry*, Springer-Verlag, New York, 1983, DOI 10.1007/978-1-4757-1810-2. MR715605

[Lev08]   A. Levin, *Difference algebra*, Algebra and Applications, vol. 8, Springer, New York, 2008, DOI 10.1007/978-1-4020-6947-5. MR2428839

[Mal01]   B. Malgrange, *Le groupoïde de Galois d'un feuilletage* (French), Essays on geometry and related topics, Vol. 1, 2, Monogr. Enseign. Math., vol. 38, Enseignement Math., Geneva, 2001, pp. 465–501. MR1929336

[Mal09]   B. Malgrange, *Pseudogroupes de lie et théorie de galois différentielle*, Preprint, 2009.

[MO11]   A. Minchenko and A. Ovchinnikov, *Zariski closures of reductive linear differential algebraic groups*, Adv. Math. **227** (2011), no. 3, 1195–1224, DOI 10.1016/j.aim.2011.03.002. MR2799605

[Mor09]   S. Morikawa, *On a general difference Galois theory. I* (English, with English and French summaries), Ann. Inst. Fourier (Grenoble) **59** (2009), no. 7, 2709–2732. MR2649331

[MU09]   S. Morikawa and H. Umemura, *On a general difference Galois theory. II* (English, with English and French summaries), Ann. Inst. Fourier (Grenoble) **59** (2009), no. 7, 2733–2771. MR2649332

[MvdP03]   B. H. Matzat and M. van der Put, *Iterative differential equations and the Abhyankar conjecture*, J. Reine Angew. Math. **557** (2003), 1–52, DOI 10.1515/crll.2003.032. MR1978401

[Ovc09a]   A. Ovchinnikov, *Differential Tannakian categories*, J. Algebra **321** (2009), no. 10, 3043–3062, DOI 10.1016/j.jalgebra.2009.02.008. MR2512641

[Ovc09b]   A. Ovchinnikov, *Tannakian categories, linear differential algebraic groups, and parametrized linear differential equations*, Transform. Groups **14** (2009), no. 1, 195–223, DOI 10.1007/s00031-008-9042-9. MR2480859

[Pil02]   A. Pillay, *Finite-dimensional differential algebraic groups and the Picard-Vessiot theory*, Differential Galois theory (Będlewo, 2001), Banach Center Publ., vol. 58, Polish Acad. Sci. Inst. Math., Warsaw, 2002, pp. 189–199, DOI 10.4064/bc58-0-14. MR1972454

[Poi90]   H. Poincaré, *Sur une classe nouvelle de transcendantes uniformes*, Journal de Mathématiques Pures et Appliquées **6** (1890), 313–365.

[Pra83]   C. Praagman, *The formal classification of linear difference operators*, Nederl. Akad. Wetensch. Indag. Math. **45** (1983), no. 2, 249–261. MR705431

[Sau00]   J. Sauloy, *Systèmes aux q-différences singuliers réguliers: classification, matrice de connexion et monodromie* (French, with English and French summaries), Ann. Inst. Fourier (Grenoble) **50** (2000), no. 4, 1021–1071. MR1799737

[Sau04a]   J. Sauloy, *Galois theory of Fuchsian q-difference equations*, Annales Scientifiques de l'École Normale Supérieure. Quatrième Série **36** (2004), no. 6, 925–968.

[Sau04b]   J. Sauloy, *La filtration canonique par les pentes d'un module aux q-différences et le gradué associé* (French, with English and French summaries), Ann. Inst. Fourier (Grenoble) **54** (2004), no. 1, 181–210. MR2069126

[Ser56]   J.-P. Serre, *Géométrie algébrique et géométrie analytique* (French), Ann. Inst. Fourier (Grenoble) **6** (1955/56), 1–42. MR82175

[Ume10]   H. Umemura, *Picard-Vessiot theory in general Galois theory*, Algebraic methods in dynamical systems, Banach Center Publ., vol. 94, Polish Acad. Sci. Inst. Math., Warsaw, 2011, pp. 263–293, DOI 10.4064/bc94-0-19. MR2905461

[vdPR07]   M. van der Put and M. Reversat, *Galois theory of q-difference equations* (English, with English and French summaries), Ann. Fac. Sci. Toulouse Math. (6) **16** (2007), no. 3, 665–718. MR2379057

[vdPS97]	M. van der Put and M. F. Singer, *Galois theory of difference equations*, Lecture Notes in Mathematics, vol. 1666, Springer-Verlag, Berlin, 1997, DOI 10.1007/BFb0096118. MR1480919
[Wat79]	W. C. Waterhouse, *Introduction to affine group schemes*, Graduate Texts in Mathematics, vol. 66, Springer-Verlag, New York-Berlin, 1979. MR547117
[Zdu97]	A. Zdunik, *Harmonic measure on the Julia set for polynomial-like maps*, Invent. Math. **128** (1997), no. 2, 303–327, DOI 10.1007/s002220050142. MR1440307

Editorial Information

To be published in the *Memoirs*, a paper must be correct, new, nontrivial, and significant. Further, it must be well written and of interest to a substantial number of mathematicians. Piecemeal results, such as an inconclusive step toward an unproved major theorem or a minor variation on a known result, are in general not acceptable for publication.

Papers appearing in *Memoirs* are generally at least 80 and not more than 200 published pages in length. Papers less than 80 or more than 200 published pages require the approval of the Managing Editor of the Transactions/Memoirs Editorial Board. Published pages are the same size as those generated in the style files provided for $\mathcal{AMS}$-LaTeX.

Information on the backlog for this journal can be found on the AMS website starting from http://www.ams.org/memo.

A Consent to Publish is required before we can begin processing your paper. After a paper is accepted for publication, the Providence office will send a Consent to Publish and Copyright Agreement to all authors of the paper. By submitting a paper to the *Memoirs*, authors certify that the results have not been submitted to nor are they under consideration for publication by another journal, conference proceedings, or similar publication.

Information for Authors

Memoirs is an author-prepared publication. Once formatted for print and on-line publication, articles will be published as is with the addition of AMS-prepared frontmatter and backmatter. Articles are not copyedited; however, confirmation copy will be sent to the authors.

Initial submission. The AMS uses Centralized Manuscript Processing for initial submissions. Authors should submit a PDF file using the Initial Manuscript Submission form found at www.ams.org/submission/memo, or send one copy of the manuscript to the following address: Centralized Manuscript Processing, MEMOIRS OF THE AMS, 201 Charles Street, Providence, RI 02904-2294 USA. If a paper copy is being forwarded to the AMS, indicate that it is for *Memoirs* and include the name of the corresponding author, contact information such as email address or mailing address, and the name of an appropriate Editor to review the paper (see the list of Editors below).

The paper must contain a *descriptive title* and an *abstract* that summarizes the article in language suitable for workers in the general field (algebra, analysis, etc.). The *descriptive title* should be short, but informative; useless or vague phrases such as "some remarks about" or "concerning" should be avoided. The *abstract* should be at least one complete sentence, and at most 300 words. Included with the footnotes to the paper should be the 2020 *Mathematics Subject Classification* representing the primary and secondary subjects of the article. The classifications are accessible from www.ams.org/msc/. The Mathematics Subject Classification footnote may be followed by a list of *key words and phrases* describing the subject matter of the article and taken from it. Journal abbreviations used in bibliographies are listed in the latest *Mathematical Reviews* annual index. The series abbreviations are also accessible from www.ams.org/msnhtml/serials.pdf. To help in preparing and verifying references, the AMS offers MR Lookup, a Reference Tool for Linking, at www.ams.org/mrlookup/.

Electronically prepared manuscripts. The AMS encourages electronically prepared manuscripts, with a strong preference for $\mathcal{AMS}$-LaTeX. To this end, the Society has prepared $\mathcal{AMS}$-LaTeX author packages for each AMS publication. Author packages include instructions for preparing electronic manuscripts, samples, and a style file that generates the particular design specifications of that publication series.

Authors may retrieve an author package for *Memoirs of the AMS* from www.ams.org/journals/memo/memoauthorpac.html. The *AMS Author Handbook* is available in PDF format from the author package link. The author package can also be obtained free of charge by sending email to tech-support@ams.org or from the Publication Division,

American Mathematical Society, 201 Charles St., Providence, RI 02904-2294, USA. When requesting an author package, please specify the publication in which your paper will appear. Please be sure to include your complete mailing address.

After acceptance. The source files for the final version of the electronic manuscript should be sent to the Providence office immediately after the paper has been accepted for publication. The author should also submit a PDF of the final version of the paper to the editor, who will forward a copy to the Providence office.

Accepted electronically prepared files can be submitted via the web at **www.ams.org/submit-book-journal/**, sent via FTP, or sent on CD to the Electronic Prepress Department, American Mathematical Society, 201 Charles Street, Providence, RI 02904-2294 USA. TeX source files and graphic files can be transferred over the Internet by FTP to the Internet node **ftp.ams.org** (130.44.1.100). When sending a manuscript electronically via CD, please be sure to include a message indicating that the paper is for the *Memoirs*.

Electronic graphics. Comprehensive instructions on preparing graphics are available at **www.ams.org/authors/journals.html**. A few of the major requirements are given here.

Submit files for graphics as EPS (Encapsulated PostScript) files. This includes graphics originated via a graphics application as well as scanned photographs or other computer-generated images. If this is not possible, TIFF files are acceptable as long as they can be opened in Adobe Photoshop or Illustrator.

Authors using graphics packages for the creation of electronic art should also avoid the use of any lines thinner than 0.5 points in width. Many graphics packages allow the user to specify a "hairline" for a very thin line. Hairlines often look acceptable when proofed on a typical laser printer. However, when produced on a high-resolution laser imagesetter, hairlines become nearly invisible and will be lost entirely in the final printing process.

Screens should be set to values between 15% and 85%. Screens which fall outside of this range are too light or too dark to print correctly. Variations of screens within a graphic should be no less than 10%.

Any graphics created in color will be rendered in grayscale for the printed version unless color printing is authorized by the Managing Editor and the Publisher. In general, color graphics will appear in color in the online version.

Inquiries. Any inquiries concerning a paper that has been accepted for publication should be sent to **memo-query@ams.org** or directly to the Electronic Prepress Department, American Mathematical Society, 201 Charles St., Providence, RI 02904-2294 USA.

Editors

This journal is designed particularly for long research papers, normally at least 80 pages in length, and groups of cognate papers in pure and applied mathematics. Papers intended for publication in the *Memoirs* should be addressed to one of the following editors. The AMS uses Centralized Manuscript Processing for initial submissions to AMS journals. Authors should follow instructions listed on the Initial Submission page found at www.ams.org/memo/memosubmit.html.

Managing Editor: Henri Darmon, Department of Mathematics, McGill University, Montreal, Quebec H3A 0G4, Canada; e-mail: darmon@math.mcgill.ca

1. GEOMETRY, TOPOLOGY & LOGIC

    Coordinating Editor: Richard Canary, Department of Mathematics, University of Michigan, Ann Arbor, MI 48109-1043 USA; e-mail: canary@umich.edu

    Algebraic topology, Michael Hill, Department of Mathematics, University of California Los Angeles, Los Angeles, CA 90095 USA; e-mail: mikehill@math.ucla.edu

    Logic, Mariya Ivanova Soskova, Department of Mathematics, University of Wisconsin–Madison, Madison, WI 53706 USA; e-mail: msoskova@math.wisc.edu

    Low-dimensional topology and geometric structures, Richard Canary

    Symplectic geometry, Yael Karshon, School of Mathematical Sciences, Tel-Aviv University, Tel Aviv, Israel; and Department of Mathematics, University of Toronto, Toronto, Ontario M5S 2E4, Canada; e-mail: karshon@math.toronto.edu

2. ALGEBRA AND NUMBER THEORY

    Coordinating Editor: Henri Darmon, Department of Mathematics, McGill University, Montreal, Quebec H3A 0G4, Canada; e-mail: darmon@math.mcgill.ca

    Algebra, Radha Kessar, Department of Mathematics, City, University of London, London EC1V 0HB, United Kingdom; e-mail: radha.kessar.1@city.ac.uk

    Algebraic geometry, Lucia Caporaso, Department of Mathematics and Physics, Roma Tre University, Largo San Leonardo Murialdo, I-00146 Rome, Italy; e-mail: LCedit@mat.uniroma3.it

    Analytic number theory, Lillian B. Pierce, Department of Mathematics, Duke University, 120 Science Drive Box 90320, Durham, NC 27708 USA; e-mail: pierce@math.duke.edu

    Arithmetic geometry, Ted C. Chinburg, Department of Mathematics, University of Pennsylvania, Philadelphia, PA 19104-6395 USA; e-mail: ted@math.upenn.edu

    Commutative algebra, Irena Peeva, Department of Mathematics, Cornell University, Ithaca, NY 14853 USA; e-mail: irena@math.cornell.edu

    Number theory, Henri Darmon

3. GEOMETRIC ANALYSIS & PDE

    Coordinating Editor: Alexander A. Kiselev, Department of Mathematics, Duke University, 120 Science Drive, Rm 117 Physics Bldg, Durham, NC 27708 USA; e-mail: kiselev@math.duke.edu

    Differential geometry and geometric analysis, Ailana M. Fraser, Department of Mathematics, University of British Columbia, 1984 Mathematics Road, Room 121, Vancouver BC V6T 1Z2, Canada; e-mail: afraser@math.ubc.ca

    Harmonic analysis and partial differential equations, Monica Visan, Department of Mathematics, University of California Los Angeles, 520 Portola Plaza, Los Angeles, CA 90095 USA; e-mail: visan@math.ucla.edu

    Partial differential equations and functional analysis, Alexander A. Kiselev

    Real analysis and partial differential equations, Joachim Krieger, Bâtiment de Mathématiques, École Polytechnique Fédérale de Lausanne, Station 8, 1015 Lausanne Vaud, Switzerland; e-mail: joachim.krieger@epfl.ch

4. ERGODIC THEORY, DYNAMICAL SYSTEMS & COMBINATORICS

    Coordinating Editor: Vitaly Bergelson, Department of Mathematics, Ohio State University, 231 W. 18th Avenue, Columbus, OH 43210 USA; e-mail: vitaly@math.ohio-state.edu

    Algebraic and enumerative combinatorics, Jim Haglund, Department of Mathematics, University of Pennsylvania, Philadelphia, PA 19104 USA; e-mail: jhaglund@math.upenn.edu

    Probability theory, Robin Pemantle, Department of Mathematics, University of Pennsylvania, 209 S. 33rd Street, Philadelphia, PA 19104 USA; e-mail: pemantle@math.upenn.edu

    Dynamical systems and ergodic theory, Ian Melbourne, Mathematics Institute, University of Warwick, Coventry CV4 7AL, United Kingdom; e-mail: I.Melbourne@warwick.ac.uk

    Ergodic theory and combinatorics, Vitaly Bergelson

5. ANALYSIS, LIE THEORY & PROBABILITY

    Coordinating Editor: Stefaan Vaes, Department of Mathematics, Katholieke Universiteit Leuven, Celestijnenlaan 200B, B-3001 Leuven, Belgium; e-mail: stefaan.vaes@wis.kuleuven.be

    Functional analysis and operator algebras, Stefaan Vaes

    Harmonic analysis, PDEs, and geometric measure theory, Svitlana Mayboroda, School of Mathematics, University of Minnesota, 206 Church Street SE, 127 Vincent Hall, Minneapolis, MN 55455 USA; e-mail: svitlana@math.umn.edu

    Probability theory and stochastic analysis, Davar Khoshnevisan, Department of Mathematics, The University of Utah, Salt Lake City, UT 84112 USA; e-mail: davar@math.utah.edu

SELECTED PUBLISHED TITLES IN THIS SERIES

1371 **H. Flenner, S. Kaliman, and M. Zaidenberg,** Cancellation for surfaces revisited, 2022

1370 **Michele D'Adderio, Alessandro Iraci, and Anna Vanden Wyngaerd,** Decorated Dyck Paths, Polyominoes, and the Delta Conjecture, 2022

1369 **Stefano Burzio and Joachim Krieger,** Type II blow up solutions with optimal stability properties for the critical focussing nonlinear wave equation on $\mathbb{R}^{3+1}$, 2022

1368 **Dounnu Sasaki,** Subset currents on surfaces, 2022

1367 **Mark Gross, Paul Hacking, and Bernd Siebert,** Theta Functions on Varieties with Effective Anti-Canonical Class, 2022

1366 **Miki Hirano, Taku Ishii, and Tadashi Miyazaki,** Archimedean Zeta Integrals for $GL(3) \times GL(2)$, 2022

1365 **Alessandro Andretta and Luca Motto Ros,** Souslin Quasi-Orders and Bi-Embeddability of Uncountable Structures, 2022

1364 **Marco De Renzi,** Non-Semisimple Extended Topological Quantum Field Theories, 2022

1363 **Alan Hammond,** Brownian Regularity for the Airy Line Ensemble, and Multi-Polymer Watermelons in Brownian Last Passage Percolation, 2022

1362 **John Voight and David Zureick-Brown,** The Canonical Ring of a Stacky Curve, 2022

1361 **Nuno Freitas and Alain Kraus,** On the Symplectic Type of Isomorphisms of the p-Torsion of Elliptic Curves, 2022

1360 **Alexander V. Kolesnikov and Emanuel Milman,** Local L^p-Brunn-Minkowski Inequalities for $p < 1$, 2022

1359 **Franck Barthe and Paweł Wolff,** Positive Gaussian Kernels Also Have Gaussian Minimizers, 2022

1358 **Swee Hong Chan and Lionel Levine,** Abelian Networks IV. Dynamics of Nonhalting Networks, 2022

1357 **Camille Laurent and Matthieu Léautaud,** Tunneling Estimates and Approximate Controllability for Hypoelliptic Equations, 2022

1356 **Matthias Grüninger,** Cubic Action of a Rank One Group, 2022

1355 **David A. Craven,** Maximal PSL_2 Subgroups of Exceptional Groups of Lie Type, 2022

1354 **Gian Paolo Leonardi, Manuel Ritoré, and Efstratios Vernadakis,** Isoperimetric Inequalities in Unbounded Convex Bodies, 2022

1353 **Clifton Cunningham, Andrew Fiori, Ahmed Moussaoui, James Mracek, and Bin Xu,** Arthur Packets for p-adic Groups by Way of Microlocal Vanishing Cycles of Perverse Sheaves, with Examples, 2022

1352 **Bernhard Mühlherr, Richard Weiss, and Holger P. Petersson,** Tits Polygons, 2022

1351 **Athanassios S. Fokas and Jonatan Lenells,** On the Asymptotics to all Orders of the Riemann Zeta Function and of a Two-Parameter Generalization of the Riemann Zeta Function, 2022

1350 **Çağatay Kutluhan, Steven Sivek, and C. H. Taubes,** Sutured ECH is a Natural Invariant, 2022

1349 **Leonard Gross,** The Yang-Mills Heat Equation with Finite Action in Three Dimensions, 2022

1348 **Murat Akman, Jasun Gong, Jay Hineman, John Lewis, and Andrew Vogel,** The Brunn-Minkowski Inequality and a Minkowski Problem for Nonlinear Capacity, 2022

1347 **Zhiwu Lin and Chongchun Zeng,** Instability, Index Theorem, and Exponential Trichotomy for Linear Hamiltonian PDEs, 2022

For a complete list of titles in this series, visit the
AMS Bookstore at **www.ams.org/bookstore/memoseries/**.